Rhizosphere
Dynamics

AAAS Selected Symposia Series

Rhizosphere Dynamics

Edited by James E. Box, Jr.,
and Luther C. Hammond

Routledge
Taylor & Francis Group

LONDON AND NEW YORK

AAAS Selected Symposium **113**

First published 1990 by Westview Press

Published 2018 by Routledge
52 Vanderbilt Avenue, New York, NY 10017
2 Park Square, Milton Park, Abingdon, Oxon OX14 4RN

Routledge is an imprint of the Taylor & Francis Group, an informa business

Library of Congress Cataloging-in-Publication Data
Rhizosphere dynamics / edited by James E. Box, Jr., and Luther
  C. Hammond
    p. cm. — (AAAS selected symposia series ; #113)
  Includes bibliographical references.
  ISBN 0-8133-7955-5
  1. Soil ecology. 2. Soil biology. 3. Roots (Botany)—Ecology.
I. Box, James E., 1931–   . II. Hammond, Luther C., 1921–
III. Series: AAAS selected symposium ; 113.
QH541.5.S6R48 1990
574.5′26404–dc20   90-32324
                                                                         CIP

ISBN 13: 978-0-367-28604-0 (hbk)

ISBN 13: 978-0-367-30150-7 (pbk)

ABOUT THE SERIES

The AAAS Selected Symposia Series was begun in 1977 to provide a means of more permanently recording and more widely disseminating some of the valuable material which is discussed at the AAAS Annual National Meetings. The volumes in the Series are based on symposia held at the Meetings which address topics of current and continuing significance, both within and among the sciences, and in the areas in which science and technology have an impact on public policy. The Series format is designed to provide for rapid dissemination of information, so the papers are reproduced directly from camera-ready copy. The papers are organized and edited by the symposium arrangers who then become the editors of the various volumes. Most papers published in the Series are original contributions which have not been previously published, although in some cases additional papers from other sources have been added by an editor to provide a more comprehensive view of a particular topic. Symposia may be reports of new research or reviews of established work, particularly work of an interdisciplinary nature, since the AAAS Annual Meetings typically embrace the full range of the sciences and their societal implications.

ARTHUR HERSCHMAN
Head, Meetings and Publications
American Association for the
Advancement of Science

CONTENTS

List of Tables and Figures...............................xi
Acknowledgment...xix

Chapter 1

INTRODUCTION
  J. E. Box, Jr. and L. C. Hammond.......................1

References..6

Chapter 2

SOIL STRUCTURAL INFLUENCES ON THE ROOT
  ZONE AND RHIZOSPHERE
  R. R. Allmaras and S. D. Logsdon.......................8

Spatial Arrangement of Soil Structure
  in Arable Agriculture..................................9
Soil Structural Influences on Root Zone
  and Rhizosphere Environment...........................15
Root Development in the Spatially Variable
  Root Environment.....................................33
References...45

Chapter 3

SOIL ORGANIC MATTER, TILLAGE, AND THE RHIZOSPHERE
  C. E. Clapp, J. A. E. Molina, and R. H. Dowdy.........55

Soil Organic Matter...................................55
Carbon and Nitrogen Cycles in the
  Rhizosphere...66

Soil Tillage and Management............................71
Summary...73
References..75

Chapter 4

TOOLS FOR STUDYING RHIZOSPHERE DYNAMICS
    Dan R. Upchurch and Howard M. Taylor..................83

Destructive Techniques...............................86
Nondestructive Techniques............................97
Root System Parameters..............................110
Conclusions...110
References..111

Chapter 5

MICROBIAL RESPONSES IN THE RHIZOSPHERE OF
    AGRICULTURAL PLANTS
    P. G. Hunt..116

Rhizosphere Establishment and Soil Structure..........117
Associative Nitrogen Fixation.........................119
Symbiotic Nitrogen Fixation...........................122
Summary...131
References..135

Chapter 6

DISTRIBUTION, STRUCTURE, AND FUNCTION OF EXTERNAL
    HYPHAE OF VESICULAR-ARBUSCULAR MYCORRHIZAL FUNGI
    David M. Sylvia......................................144

Distribution of External Hyphae.......................145
Morphology and Development of External Hyphae.........152
Function of Hyphae in Soil............................153
Methods of Quantification.............................158
References..160

Chapter 7

COLLEMBOLAN POPULATIONS AND ROOT DYNAMICS
    IN MICHIGAN AGROECOSYSTEMS
    R. J. Snider, R. Snider, and A. J. M. Smucker........168

Minirhizotron and Video Images of the
    Rhizosphere..168

Extraction of Collembola from Soil Cores...............172
Discussion...187
References...190

Chapter 8

EARTHWORMS AND AGRICULTURAL MANAGEMENT
    E. J. Kladivko and H. J. Timmenga 192

Basic Ecology of Earthworms............................192
The Effects of Earthworms on Soil Properties
    and Processes......................................194
The Effects of Earthworms on Plant Growth
    and Yield..199
Effect of Agricultural Management on
    Earthworms...204
Summary and Implications...............................207
References...210

Chapter 9

SHOOT/ROOT RELATIONSHIPS AND BIOREGULATION
    M. J. Kasperbauer...................................217

Genetic Control..218
Environmentally Induced Regulation.....................219
Acknowledgments..229
References...230

Chapter 10

ROOT-SHOOT RELATIONSHIPS IN COTTON
    B. L. McMichael.....................................232

Shoot Development......................................233
Development of the Root System.........................234
Measurement of Root-Shoot Ratios in Cotton.............237
Dynamics of Root-Shoot Relationships...................240
Factors Influencing Root-Shoot Relationships
    in Cotton..242
Genetic Aspects of Root-Shoot Relationships............245
Conclusion...246
References...247

Chapter 11

THE EFFECTS OF GRAVITY ON THE ECOLOGY
AND DYNAMICS OF ROOT GROWTH
Randy Moore..252

Influence of Gravity on Growth and
   Initiation of Roots..............................253
Influence of Gravity on Structure and
   Function of Cells in Roots.......................254
Future Prospects.....................................257
References...258

Chapter 12

EFFECTS OF MECHANICAL IMPEDANCE
ON ROOT DEVELOPMENT
M. C. Whalen and L. J. Feldman.......................260

References...266

Chapter 13

SOIL AND PLANT ROOT WATER FLUX IN THE RHIZOSPHERE
Morris G. Huck and Gerrit Hoogenboom.................268

Soil Water Storage...................................269
Soil Water Movement..................................271
Soil-Root-Water Interactions.........................278
Water Uptake by Roots................................280
Root and Soil Water Modeling.........................285
Summary..300
References...303

Index..313

About the Contributors...............................317

TABLES AND FIGURES

TABLES

2.1   Typical variations of soil structure in
      the various root-zone layers as observed
      pedologically.......................................18

2.2   Components of energy balance during the
      positive net radiation period for tillage
      treatments shown in Figure 2.3....................26

3.1   Organic matter and nitrogen contents of
      mineral surface soils from several U.S.
      locations...57

3.2   Distribution of organic matter in four
      soil profiles.....................................58

3.3   Fractionation of soil nitrogen following
      acid hydrolysis...................................60

3.4   Effect of cultivation and cropping system
      on distribution of the forms of nitrogen in
      soil..61

3.5   Carbon, oxygen, hydrogen, and nitrogen
      contents of humic acids (HA) and fulvic
      acids (FA) isolated by various solvents...........64

3.6   Carbon input from roots...........................68

TABLES

5.1   Nodular occupancy by eight Bradyrhizobium
        japonicum strains as affected by soybean
        cultivar, tillage, and inoculation with
        strain 110 of B. japonicum........................126

5.2   Two-year mean seed yield and seed yield
        rank for soybean inoculated with different
        B. japonicum strain and grown under irrigated
        (+) and nonirrigated (-) conditions in 1981
        and 1983...132

5.3   Effects of red (R) and (far-red) FR light
        treatment of shoots and root inoculation
        time on nodule number for soybean grown in
        a split-root system..............................133

7.1.   Total number of Folsomia candida in
        each cm3 of soil as measured by the
        heat extraction method, September 1986...........177

7.2.   Regression correlations between the
        numbers of Collembola and root densities
        at 10-cm intervals to a soil depth of
        100 cm for a period of 35 d in a soybean
        field..184

7.3.   Average collembolan and root distribution
        in a Metea loam soil containing soybeans
        (Glycine max, L.) and corn (Zea mays, L.)
        during the vegetative growth stages..............187

8.1.   The number of burrows open to the soil
        surface after six weeks of earthworm
        activity in Minnesota, as affected by
        tillage treatment and species....................195

8.2.   Root growth of barley grown in
        undisturbed profiles of silty clay
        loam soil, as affected by earthworm
        additions and tillage treatment..................201

8.3.   Pasture yields in New Zealand, as
        affected by earthworm (A. caliginosa)
        introduction.....................................202

TABLES

8.4. Earthworm populations as affected by
      crop and tillage practice, autumn 1983,
      in Indiana...206

9.1. Influence of FR/R ratio on soybean
      shoot/root relationships.........................226

10.1 Dry weight, starch concentrations, and
      root-shoot relationships between an annual
      and perennial variety of cotton and at two
      harvest dates....................................241

FIGURES

2.1 Model of rooted zone in arable agriculture.........11

2.2 Typical changes in the soil water content
     of the soil profile as related to the
     management of tillage and residue cover...........22

2.3 Typical influence of tillage and straw
     mulch on the diurnal variation of soil
     temperature in the tilled layer...................25

2.4 D/D_0 for H_2 as a function of ξ that was
     varied in a bed of aggregates by wetting
     or drying and also by compaction..................30

2.5 Porosity components as a function of
     water content in a Nicollet clay loam
     after different uniaxial stresses were
     imposed...32

2.6 Specific corn root length at sixth leaf
     stage for four selected interrow variants
     in 1968...35

2.7 Mapping on a horizontal plane of a compact
     treatment at 17 cm depth..........................40

2.8 Vertical mapping of a compact treatment
     showing a "shadow zone" caused by a structural
     obstacle within the plowed layer..................41

FIGURES

2.9   Corn root length density with depth at two
      dates showing the development of compensatory
      root growth deeper in the soil when the
      surface soil layers have dried.....................43

4.1   Cotton root system excavated from the top
      200 mm of the soil profile........................89

4.2   Framed monolith being extracted using the
      procedure of Nelson and Allmaras..................90

4.3   Trench profile wall with grid in place............93

4.4   Intact soil core, 76 mm in diameter,
      collected with a steel sampling tube
      hydraulically driven into the soil................94

4.5   Typical rhizotron facility, showing the
      underground walkway lined with clear
      windows through which roots are observed...........99

4.6   Minirhizotron system being used in a cotton
      field..102

4.7   Hydroponic system in which plants are held
      in styrofoam blocks and roots are suspended
      in an aerated nutrient solution contained
      in plastic cylinders.............................106

4.8   Slant box system used in greenhouse studies......108

5.1   A schematic of root development, rhizo-
      sphere establishment, and shoot-root-
      microbial equilibrium............................118

5.2   Coker 338 soybean yield response to row
      orientation, inoculation, and irrigation
      in 1982..131

6.1   Proportion of viable to total external
      hyphal length of Glomus clarum inoculated
      on Trifolium repens in three experiments.........147

FIGURES

6.2   Total and active hyphae and root coloniza-
       tion in pot cultures of Glomus spp. grown
       with Paspalum notatum for 13 weeks.................148

6.3   Effect of applied phosphorus on length of
       VAM hyphae in soil................................149

6.4   Production of hyphae of Glomus fasciculatum
       (FAS), G. tenue (TEN), Gigaspora calospora
       (CAL), and Acaulospora laevis (LEA) in soil.......150

6.5   Relationship between plant growth and pro-
       duction of external hyphae by VAM fungi...........151

6.6 ^{32}P present in root segments from mycor-
       rhizal and nonmycorrhizal onions as a func-
       tion of distance from tracer injection point
       to root surface...................................154

6.7   Water conductivity, units for flow rate are
       cm^3 sec^{-1} cm^{-1} x 10^{-8}, of roots of
       mycorrhizal and nonmycorrhizal red clover
       plants grown at three phosphorus levels...........156

6.8   Relationship between hyphal length in soil
       and percentage of water-stable aggregates
       greater than 2 mm..............................157

7.1   Video images from minirhizotron tubes.............171

7.2   Collembolan population profiles in a sugar-
       beet and Metea soil agroecosystem during
       the production season of 1986..................173

7.3   Collembolan population profiles in a corn
       and Metea soil agroecosystem during the
       production season of 1985.......................174

7.4   Collembolan population profiles in a soy-
       bean and Metea soil agroecosystem during
       the production season of 1985..................175

7.5   Collembolan population profiles in a soy-
       bean and Metea soil agroecosystem during
       the autumn and winter of 1985 and 1986........176

FIGURES

7.6   Collembolan population profiles in a soy-
      bean and Capac soil agroecosystem during
      the production season of 1985....................178

7.7   Collembolan population profiles in a corn
      and Capac soil agroecosystem during the
      production season of 1985........................179

7.8   Root development profiles of soybeans in
      a Metea soil, 1986...............................180

7.9   Root development profiles of corn in a
      Metea soil, 1986.................................181

7.10 Root development profiles for sugarbeets
      in a Metea soil, 1986............................182

7.11 Root activity profiles for soybeans in a
      Metea soil, 1986.................................183

7.12 Root activity profiles of corn in a Metea
      soil, 1986.......................................185

7.13 Root activity profiles for sugarbeets in a
      Metea soil, 1986.................................186

8.1   Corn roots growing in earthworm burrows at
      a depth of 60 cm.................................200

9.1   Influence of nitrogen deficiency on seedling
      shoot/root relationships.........................220

9.2   Field-grown biennial sweetclover tap roots
      dug August 20, September 20, October 20,
      and November 20..................................222

9.3   Tap roots from first year biennial sweet-
      clover plants grown on natural photoperiods
      with natural and 22°C minimum temperatures
      until November...................................223

9.4   Plant size and flowering condition and
      root size of biennial sweet-clover plants
      after 100 days of exposure to photoperiod
      treatment..224

FIGURES

9.5   Absorption, transmission, and reflection of
        light from a typical soybean leaf.................227

10.1   Distribution of dry matter in 'Acala 1517-C'
        cotton as a function of days after emergence......235

10.2   Development of plant height, leaf area,
        root length of field grown 'Acala SJ-5'
        cotton grown under dryland conditions.............236

10.3   Development of shoot weight, root weight,
        fruit weight, and root-shoot ratios in
        greenhouse-grown "Paymaster 145" cotton...........238

10.4   Partitioning of fresh and dry weights of
        organs of 'Acala SJ-2' cotton, on a per/
        plant basis, grown in pots of different
        capacities..239

13.1   Soil particles surrounded by films of
        capillary water and connecting irregular
        pore spaces......................................272

13.2   Forrester flow diagram for water balance
        algorithm...275

13.3   System boundaries for water movement in a
        soil with an active root system (hypothetical
        soybean plant growing in a one-dimensional
        layered soil consisting of infinite uniform
        layers in the horizontal direction)...............287

13.4   Forrester flow diagram for carbon balance
        algorithm...288

13.5   Simulated shoot and root growth rates for
        irrigated or non-irrigated plants.................290

13.6   Simulated root length from calendar day
        150 to 250 for depths indicated...................292

13.7   Simulated root growth rates from calendar
        day 200 to 210 for depths indicated...............293

FIGURES

13.8   Vertical root distribution profile simulated
       for days 200, 205, and 210......................295

13.9   Simulated soil water potential at a
       depth of 0.4 m, compared with tensiometer
       measurement and rainfall data...................296

13.10 Vertical water content profile simulated
       for days 200, 205, and 210......................298

13.11 Simulated water uptake rates from calendar
       day 200 to 210 for roots growing at the depth
       and treatment indicated.........................299

13.12 Vertical water uptake profiles computed at
       ten-day intervals from day 150 to day 250
       for each treatment..............................301

ACKNOWLEDGMENT

The editors recognize and thank Ms. Debbie Perry, Editorial Assistant, who did the word processing of the final copy, for her many hours of patient work and attention to detail.

James E. Box, Jr.
Luther C. Hammond

J. E. Box, Jr.
L. C. Hammond

1 Introduction

> Nothing in the world will take the
> place of persistence. Talent will
> not; nothing is more common than unsuc-
> cessful men with talent. Genius will
> not; unrewarded genius is almost a
> proverb. Education will not; the world
> is full of educated derelicts. Persis-
> tence and determination alone are
> omnipotent. The slogan "Press on" has
> solved and always will solve the prob-
> lems of the human race.
>
> --Calvin Coolidge

Nowhere is this statement more applicable than in the efforts of scientists to sort out the complexities of that portion of the soil profile occupied by roots. The purpose of this treatise is to explore some of these complexities with emphasis on bringing together interests of classical soil biologists and classical soil scientists. We view this "bringing together" as the essence of "Soil Biology."

This book had its beginning in the 1986 meeting of the Southern Regional Information Exchange Group, "The Plant Root Environment," in Gainesville, Florida. Subsequently, a symposium on rhizosphere dynamics was organized and presented at the 1987 AAAS Annual Meeting in Chicago. It is noteworthy that an earlier book, "The Plant Root and Its Environment" (Carson 1974), was written by forerunners of the current Southern Regional Information Exchange Group. In fact, several authors were contributors to both books.

Chapters in this treatise were written independently.

Consequently, there is some overlapping in subject matter treatment, as well as an inevitable paucity of discussion in other areas. Soil chemistry and plant nutrition were not assigned topics in contrast to the rather extensive treatment of these and associated topics in the 1964 book.

Four chapters (5 through 8) deal with soil biology, an area often ignored by scientists studying soil physical, chemical, and mineralogical factors in plant growth. All too frequently, however, microbiologist have concentrated in the small zone around the root that, for nearly 100 years, they have called the rhizosphere. The rhizosphere was first defined by Hiltner (1904) as the zone of stimulated bacterial growth around the roots of legumes. The stimulated growth was thought to be the result of nitrogen compounds released from roots. Bowen (1980) states that "...distribution of roots and microbial movement to them are essential considerations in understanding the composition of the root microflora." Hiltner's description has been modified to include all microbial growth using root-derived compounds as sources of carbon, nitrogen, and energy. Further delineation of the definition has been suggested that includes endorhizosphere, the root epidermis and cortex zone (Balanderu and Knowles 1978), and ectorhizosphere (Lynch 1982), the zone colonized outside the root and including the surrounding soil. We suggest that the rhizosphere be defined as the whole soil mass occupied by roots.

Once a root occupies a volume of the soil, it affects and is affected by the microorganisms and mesofauna of that region. Thus, the root has an immediate effect on ecology of the soil. Roots excrete large amounts of photosynthate (Lynch et al. 1981), and these organic compounds serve as substrate for microbial populations. Root colonization by vesicular-arbuscular mycorrhizal fungi (VAM) often leads to improved plant growth in nutrient-poor soils, primarily because of increased uptake of phosphorus and other poorly mobile ions. Mycorrhizal plants are frequently more tolerant of water stress, pathogens, salt, and heavy metal than nonmycorrhizal plants. The external phase of VAM fungal growth has not received adequate study. The current knowledge of the distribution, structure, and function of the external hyphae of VAM fungi; current techniques for assessing development of VAM hyphae in soil; and research needs are suggested.

The ecological significance of soil mesofauna has remained obscure for many years. Collembola are known to be significant regulators of the decomposition of soil organic matter. They have also been shown to directly affect plant growth in agroecosystems through mycorrhizal grazing (Warnock et al. 1982) and direct root consumption (Brown 1985). Earthworms can have a major impact on plant rooting in the rhizosphere by their feeding, castings, and burrowing activities. No-tillage agricultural soils may be so permeated with burrows that surface water infiltrates rapidly and no surface runoff occurs. The adverse affect can be ground water pollution from surface-applied chemicals (Edwards et al. 1979). These same burrows, however, may serve as channels for plant roots and provide nutrients from the organic coatings left by worms (Edwards and Lofty 1978). Thus, continual and dynamic relationships exist among plants, rhizosphere microorganisms and mesofauna, and the soil environment. The new focus on Soil Biology seems to hold great promise for improving plant productivity. An excellent text exists, "Soil Biotechnology" by J. M. Lynch (1983), that emphasizes applying soil microbiology "practically in a research and advisory capacity." Advancing our understanding of the complex interactions among the physical, chemical, environmental, and biological components of the natural soil system, in which farmers produce the plants that provide food we eat, is going to take a very large measure of "persistence and determination."

Soil physical factors and their influences on root development and growth provide the primary focus of Chapters 2, 3, and 13. Soil structure is characterized as the major indirect and direct physical factor affecting the soil environment and root proliferation in soil material. Yet the authors of these chapters encountered an all-too-common problem with the term "soil structure" as currently defined in the United States, "the aggregation of primary soil particles into compound particles, or clusters of primary particles, which are separated from adjoining aggregates by surfaces of weakness" (Soil Survey Staff 1975). The emphasis is on the aggregates of peds to the exclusion of consideration of the unaggregated primary particles and the voids. Furthermore, one must deal with the dilemma of characterizing the structure of sands.

A simple and more inclusive definition of soil structure has been proposed by an Australian soil scientist: "The physical constitution of a soil material as expressed by the size, shape, and arrangement of the solid particles and voids, including both the primary particles to form compound particles and the compound particles themselves" (Brewer 1964; Brewer and Sleeman 1988). This concept of soil structure should help the reader gain additional insights on the importance of soil structure and its interre-

lations with other physical factors, tillage, organic matter, and root growth. Organic matter in surface soils
exhibits both direct and indirect effects on the rhizosphere. The amount and distribution, composition, and functions of the nonliving component of soil organic matter is
important to biological activity in the root-zone. Amounts
and distribution of soil organic matter are regulated by
the so-called "factors of soil formation"--time, climate,
vegetation, parent material, and topography (Jenny 1941).

The most universal limiting factor for plant root
growth is water. Most frequently there is not enough water, but occasionally there is too much of it. Understanding the flow of water into and through the soil and to the
root surface is critical. Our final chapter provides a
rather extensive treatment of soil water. Water entry and
movement in the soil is a complex process that depends upon
soil physical properties, including boundary conditions
created by an abrupt change in the continuity of pore sized
with depth across soil layers. A rate of water input to
the soil surface greater than the rate of entry through the
surface will result in either ponding or surface runoff.
When water infiltration through the soil surface ceases,
there is a variable time of redistribution of water in the
soil profile and possibly drainage from it to ground water. Usually after 2 or 3 days, the rate of redistribution
becomes slow relative to the rate of evaporation from soil
and plant surfaces, and the soil profile of specified depth
and boundary conditions is said to be at "field capacity,"
or an upper limit of water retention against drainage.
Unfortunately, the definition of field capacity has been
limited to well drained soils. It can and should be applied to poorly drained soils as well. Water which remains
in the soil profile as stored water is available to roots,
but in a complex and varied manner. These relationships
that apply to the root and soil water generally apply to
all of soil biology.

Plant physiological aspects of root dynamics are
considered in Chapters 9, 10, 11, and 12. Bioregulatory
responses to climate, light quality, soil environment, and
utilization of the above-ground portions of the plant frequently establish a constant vegetative mass relationship
between the shoots and roots. Drought or overgrazing will
reduce both above- and below-ground plant mass. Responses
to the environment are complex to the extent that light
quality (Kasperbauer and Karlen 1986) and genetics
(McMichael and Quisenberry 1986) can affect the root/shoot
ratio. The role of gravity in plant rooting has not been

understood clearly, principally because it had not been possible to design a study that completely eliminates the effect of gravity. Recent experiments have involved launching imbibed seeds and seedlings into outer space, allowing them to grow for several days, and then fixing them before re-entry into Earth's gravity (Moore et al. 1987). Gravity strongly affects growth and initiation of roots. If gravity is a subtle influence of the physical environment on plant rooting, mechanical impedance is a very unsubtle influence on rooting. Interestingly enough, both affect plant bioregulation. Since the root cap is considered to be important in the growth response of roots to mechanical impedance (Goss and Russell 1980), rates of ethylene evolution from decapped impeded roots were compared with rates from decapped control roots (Whalen 1988). It was concluded that the root cap is not required for the observed early response of ethylene metabolism to mechanical impedance.

Research on plant root systems and their associated soil microbiology and mesofauna activity is difficult under field conditions because soil limits their accessibility for observations. Many of the field methods for observing root growth and soil biology are labor intensive and require extensive destruction of the field sites (Böhm 1979). Glass wall methods were developed to observe root growth in situ so that comprehensive investigations of root ecology and function could be conducted. The modern root laboratories or rhizotrons and the glass tubes or minirhizotron systems are modifications of the glass wall methods. An overview of the historical development of the rhizotron and minirhizotron, along with some specific applications, are presented in Chapter 4.

In conclusion, the authors involved in this labor of love are to be commended for documenting some of the pieces of our rhizosphere-heritage platform, albeit an always shaky one, on which we scientists stand and peer optimistically into the future.

REFERENCES

Balandreau, J., and R. Knowles. 1978. The rhizosphere. In _Interactions between non-pathogenic soil micro-organisms and plants_, ed. Y. R. Dommergues and S. V. Krupa, 243-268. Amsterdam: Elsevier.

Böhm, W. H. 1979. _Methods for studying root systems_. New York: Springer-Verlag.

Bowen, G. D. 1980. Misconceptions, concepts and approaches in rhizosphere biology. In _Contemporary Microbial Ecology_, ed. D. C. Elwood, J. N. Hedger, M. J. Latham, J. M. Lynch, and J. H. Slater, 283-304. London: Academic Press.

Brewer, R. 1964. _Fabric and mineral analysis of soils_. New York: John Wiley & Sons, Inc.

Brewer, R., and J. R. Sleeman. 1988. _Soil structure and fabric_. East Melbourne, Australia: CSIRO.

Brown, R. A. 1985. Effects of some root-grazing arthropods on the growth of sugarbeets. In _Ecological interactions in soil_, ed. A. M. Fitter, D. Atkinson, D. Y. Read, and M. B. Usher, 285-295. Oxford, England: Blackwell Science Publishers.

Carson, E. W., ed. 1974. _The plant root and its environment_. Charlottesville: University of Virginia Press.

Edwards, C. A., and J. R. Lofty. 1978. The influence of arthropods and earthworms upon root growth of direct drilled cereals. _Journal of Applied Ecology_ 15:789-795.

Edwards, W. M., R. R. van der Ploeg, and W. Ehlers. 1979. A numerical study of the effects of noncapillary-sized pores upon infiltration. _Soil Science Society of America Journal_ 43:851-856.

Goss, M. J., and R. S. Russell. 1980. Effects of mechanical impedance on root growth in barley (_Hordeum vulgare_ L.). III. Observation on the mechanism of response. _Journal of Experimental Botany_ 31:577-588.

Hiltner, L. 1904. Uber neuere Erfahrungen und Probleme auf dem Gebiet der Bodenbakteriologie und unter besonderer Berucksichtigung der Frundungung undnBrache. _Arbeiten der Deutschen Landwirtschaftsgesellschaft_. 98:59-78.

Jenny, H. 1941. _Factors of soil formation_. New York: McGraw-Hill.

Kasperbauer, M. J., and D. L. Karlen. 1986. Light-mediated bioregulation of tillering and photosynthate partitioning in wheat. _Physiologia Plantarum_ 66:159-163.

Lynch, J. M., J. H. Slater, J. A. Bennett, and S. H. T. Harper. 1981. Cellulase activities of some aerobic micro-organisms isolated from soil. _Journal of General Microbiology_ 127:231-236.

Lynch, J. M. 1982. The rhizosphere. In _Experimental microbial ecology_, ed. R. G. Burns and S. H. Slater, 1-23. Oxford, England: Blackwell Scientific Publications.

Lynch, J. M. 1983. _Soil biotechnology_. Oxford, England: Blackwell Scientific Publications.

McMichael, B. L., and J. E. Quisenberry. 1986. Variability in lateral root development and branching intensity in exotic cotton. In _Beltwide cotton production research conference proceedings_, ed. T. C. Nelson. 89. Memphis: National Cotton Council of America.

Moore, R., W. M. Fondren, C. E. McClelen, and C-L. Wang. 1987. Influence of microgravity on cellular differentiation in root caps of _Zea mays_. _American Journal of Botany_ 74:1006-1012.

Soil Survey Staff. 1975. _Soil taxomony: A basic system of soil classification for making and interpreting soil surveys_. Agricultural Handbook no. 436, USDA, Soil Conservation Service, Washington, DC. 754 pp.

Warnock, A. J., A. H. Fitter, and M. B. Usher. 1982. The influence of a springtail _Folsomia candida_ (Insecta, Collembola) on the mycorrhizal association of leek, _Allium porrum_ and the vesicular-arbuscular mycorrhizal endophyte _Glomus fasciculatum_. _New Phytologist_ 90:285-292.

Whalen, M. C. 1988. The effect of mechanical impedance on ethylene production by maize roots. _Canadian Journal of Botany_ 66:2139-2142.

R. R. Allmaras
S. D. Logsdon

2 Soil Structural Influences on the Root Zone and Rhizosphere

The geometry of planting most annual crops may persist for at least sixty days after planting. Remote sensing measurements of reflectance taken from the nadir position (directly overhead) above these planted fields usually have a signature (a two-dimensional pattern) showing an influence of the soil surface in the interrow, while plants may dominate the signature over the row (Gardner and Blad 1986). Although these measurements from the nadir position are indicative of reflected radiation, they demonstrate a smaller leaf area index in the interrow than in the row. The extinction of short-wave radiation is less in the interrow (Norman 1979). The plane or layer for radiation exchange nearer to the soil surface in the interrow therefore produces a nonuniform thermal environment between interrow and row. Measurements of wind movement in the direction of the row in a soybean canopy (Perrier et al. 1970) also demonstrate a nonuniform thermal environment; soil evaporation was also shown to be nonuniform. Soil cracking midway between rows (Johnson 1962) is another manifestation of evaporation and greater plant use of water in the row.

A casual observer may have no suspicion that there are vertical and horizontal nonuniformities of thermal, moisture, and gaseous environments. These nonuniform thermal, moisture, and gaseous environments are induced by canopy architecture of the row crop and by soil structure variations. Soil structure is defined as "the combination or arrangement of primary particles into secondary particles, units, or peds" (Soil Science Society of America 1987). The nonuniformity suggested for annual crops also occurs in some perennial crops; for example, alfalfa regrowth responds to compaction patterns produced by traffic during harvest and transport (Grimes, Sheesley, and Wiley 1978).

Some of the vertical soil-structure nonuniformities are natural products of the soil formation process and are formally classified as soil horizons of a soil series. Other vertical and horizontal nonuniformities are produced by the action of tillage machines, traffic associated with traction and transport during tillage and seeding, harvest equipment for grain and root crops, and soil-engaging tools used to apply fertilizers and other agricultural chemicals.

Structural nonuniformities, especially those in the upper parts of the root zone, may have significant influences on plant rooting. Because plant roots are nearly always stressed under field conditions (Russell 1977), it is a challenge to manage soil environment in the root zone so that plant roots will adequately support plant uptake of nutrients and water. A systematic approach to this soil management is to describe a simple model of the root zone so that spatial variations of soil structure are systematized in relation to field operations (especially tillage and traffic) and other soil properties. One can then examine spatial and temporal variations of a root system as the root responds to soil environment.

Some simulation models of root response to soil environment (Shaffer and Larson 1987; Whisler, Lambert, and Landivar 1982) recognize a spatial character of soil environment wherein the plant root develops in the most accessible localities that provide conditions favorable to root growth. The root is an integrated system consisting of distributed meristems and zones of cell elongation and maturation. These various parts of the root do not all require the same soil environment. For example, the different zones of the corn root system each has a distinct response to water and temperature environments controllable by tillage and crop residue management (Allmaras et al. 1987).

SPATIAL ARRANGEMENT OF SOIL STRUCTURE
IN ARABLE AGRICULTURE

A first approximation of spatially variable soil structure in arable agriculture is depicted in Figure 2.1, which shows four root zone layers and their nominal thicknesses. The rhizosphere consists of the root and a thin shell that contains root exudates and some microbes not found outside this volume of influence; however, the rhizosphere environment is affected by environment in the root zone. The tilled layer is frequently disturbed (fragmented) by till-

age and variously compacted by traffic. In fields sown to
annual crops, this disturbance usually occurs during primary tillage (first and/or deepest tillage after harvest),
secondary tillage (usually not as deep as primary tillage),
and planting. Traffic occurs during all of these operations so that recompaction may occur at various depths and
horizontal positions in the tilled layer. The most frequent pattern of compaction due to traffic is that produced
in only some interrows--the generalization in Figure 2.1
depicts compaction in alternate interrows as deep as the
thickness of the tilled layer. The packed [Pa] layer is
rarely disturbed (fragmented) by tillage machinery, but may
be subjected to a characteristic deformation, manifested as
shear in its upper part (just under the tilled layer) and
as compression in the lower part. The upper subsoil [SSu]
and lower subsoil [SSl] layers are not normally disturbed
by field operations and are distinguished into two layers
merely to emphasize that the soil-forming process may produce a genetic horizon in one of these root zones, and that
biopores are often more predominant in the upper subsoil
compared to lower subsoil layer. When there is a genetic
horizon located in the upper part of the subsoil, this root
zone layer may impart a distinct influence on root development related to such environments as water flux, penetration resistance, and adverse chemical environment.

Sometimes a genetic horizon may be located in the packed layer. One must go to the soil series level to identify
the depth and thickness of genetic horizons in the root
zone scheme of Figure 2.1 (Soil Survey Staff 1975). For
purposes of illustration the soil structural features of
several soil series will be discussed later in relation to
the scheme of Figure 2.1. When severe erosion has exposed
the subsoil, the tilled and packed layers are usually distinct from each other, and the subsoil layers (of Fig. 1.1)
may yet retain features characteristic of the original subsoil.

Soil structural features of importance are evidenced
by pedality, penetrometer resistance, macroporosity, consis-
tence, dry bulk density, air and water porosity, and micro-
morphology (McKeague et al. 1987; Hamblin 1985). Pedality
"refers to the natural organization of soil particles into
units (peds) which are separated by surfaces of weakness
that persist through more than one cycle of wetting and
drying in place" (McKeague et al. 1987). Pedality merely
delimits the broader definition of soil structure given
above. Methodolgy for examining pedality is also set forth
by the Soil Survey Staff (1951).

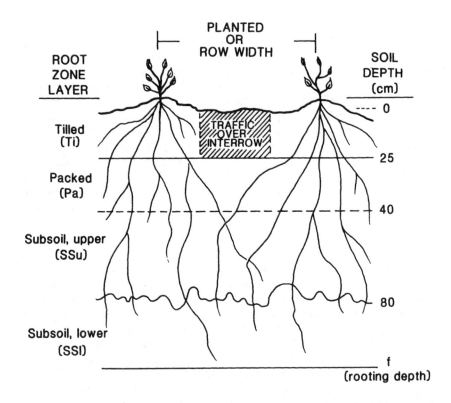

Figure 2.1. Model of rooted zone in arable agriculture.

The persistence of weakened surfaces is a factor in rooting susceptability. Rooting susceptibility can also be inferred from fluid flux rates such as saturated hydraulic conductivity (Bouma and Anderson 1973) and gaseous diffusion per unit of porosity (Douglas and Goss 1987). Even though macroporosity, including biopores, is often less than 1 percent of the pore volume, dry bulk density and resistance to penetrometer advance are not always a good indicator of rooting susceptibility (McKeague et al. 1987; Ehlers et al. 1983); however, some researchers have used dry bulk density as a function of depth to locate potential zones of soil compaction produced by traffic and tillage-tool penetration (Allmaras et al. 1988b).

The Tilled Layer in the Root Zone

The tilled layer has a common depth of 25 cm (Figure 2.1) when a moldboard plow is used for primary tillage. Dry bulk density measurements, as a function of soil depth, show a maximum immediately below the maximum penetration of the moldboard plow. Maximum tillage-tool penetration in experimental plots that had a long-term primary tillage with a moldboard plow ranged from 15 to 29 cm (Pikul and Allmaras 1986; Douglas et al. 1986; McKeague et al. 1987; Allmaras et al. 1988b). In farm fields with a long-term history of moldboard plowing, the depth of the last mold-board plow penetration ranged from 18 to 34 cm (Allmaras et al. 1988a). Organic carbon profiles in a Typic Haploxeroll with a long-term history of moldboard plowing show a pene-tration depth of 20 to 30 cm both in experimental and farm fields (Allmaras et al. 1982, 1988b).

Pedality features such as massive, granular with packing voids, or apedal with occasional firm fragments (Allmaras et al. 1982; McKeague et al. 1987; Bouma, van Rooyen, and Hole 1975) usually characterize soil structure in the tilled layer, whereas the upper part of the packed layer often shows a platy structure (Bouma, van Rooyen, and Hole 1975). Pedality features would be different in the portion of the tilled layer subjected to traffic, i.e., between, rather than in, the planted row.

Dry bulk density and organic carbon profiles (when the same tool is in long-term use) indicate that other tillage tools do not penetrate as deeply as the moldboard plow. Both profile types gave the same estimate within 3 cm depth. Penetration by a sweep was estimated to be 16 cm (Allmaras et al. 1988b); for a chisel the depth was 17 cm (Allmaras et al. 1988b); and for the spring-tine on a cultivator the penetration was 10 cm (Douglas et al. 1986). Preliminary measurements in row crops on a Typic Haplaquoll, an Aquic Hapludoll, and a Udic Haplustoll confirm that chisels and discs as primary tillage tools penetrate the soil from about 10 to 15 cm.

When chisels, spring-tine cultivators, harrows, and discs are used for secondary tillage, their penetration is usually about 10 to 15 cm. Douglas and Goss (1987) and Douglas et al. (1986) observed that even ten years after a switch from a moldboard plow to some other form of primary tillage tool, a residual pressure pan (see Soil Science Society of America 1987 for definition) is indicated by profiles of dry bulk density or penetrometer resistance. Allmaras et al. (1988a, 1988b) show that secondary tillage (with a shovel-type implement) and associated traffic compaction following moldboard plowing may only repack the soil to 15 cm, leaving a lower density layer between the recompacted layer and the maximum depth of moldboard plow penetration. In a ten-year study (Douglas and Goss 1987; Douglas et al. 1986) in which traffic was random and only a spring-tine cultivator (10 cm deep) was used for tillage, air permeability was reduced and strength was increased in the 10- to 25-cm layer when compared to soil in the 0- to 10-cm layer (and when compared with moldboard plow treatments in the 10- to 25-cm layer). Besides higher bulk density and higher penetrometer resistance (both are better measures of micro- than macroporosity), Douglas and Goss (1987) found lower macroporosity continuity at the base of the moldboard plow penetration.

The tilled layer nearly always has spatial variations of repacked soil alternating with soil that has not been packed by traffic since the last fragmentation with a primary or secondary tillage implement. Soil packed during random traffic associated with secondary tillage is usually fragmented because the depth of packing is about the same as the tool penetration, i.e. 10 to 15 cm (Allmaras, Kraft, and Miller 1988; Culley, Larson, and Randall 1987). This comparative penetration of traffic compaction and associated secondary tillage-tool is most easily seen when a moldboard plow penetrates about 25 cm for primary tillage (Allmaras et al. 1988b). On the other hand, when a disc or chisel plow is used for primary tillage (at a depth of 10 to 15 cm), the zone of secondary compaction cannot be separated from that produced during primary tillage. Soil packing after primary tillage may extend deeper than 15 cm with the compacted band nearly as thick as the tilled layer when large tractors and/or harvesters are used (Allmaras et al. 1988a; Voorhees, Nelson, and Randall 1986) or when there has been repeated traffic with no intervening cultivation as in no-till (Culley, Larson, and Randall 1987). Oriented traffic (as during planting) with no subsequent secondary tillage usually penetrates about 15 cm (Culley, Larson, and Randall 1987; Allmaras et al. 1988b). These compacted interrows can have negative impacts on crop yields. For example, Fausey and Dylla (1984) showed a negative influence on corn growth where rooting in the compacted interrow was required for nitrogen nutrition.

The Packed and Subsoil Layers in the Root Zone

The packed layer often shows evidence of a platy structure in its upper dimension (strains from tillage-tool
stresses and from compaction stresses under traffic) and an
expression of blocky, subangular blocky, or prismatic structure at the lower dimension, especially in Mollisols
(Allmaras et al. 1982; Bouma, van Rooyen, and Hole 1975;
McKeague et al. 1987). Frequently a higher bulk density
occurs in the layer with platy structure. Later it will be
noted that the soil structure description can be used to
estimate hydraulic conductivity and also rooting. In
Mollisols this layer often has evidence of biopores (as
much as 2 percent of the horizontal cross-sectional area)
occupied by roots and invertebrates. Many investigators
(Hamblin 1985; Buntley and Papendick 1960; Edwards, Norton,
and Redmond 1988; Shipitalo and Protz 1987) have noted
strong biopore character in the packed layer and biopores
extending from 70 to 150 cm deep, into the upper subsoil
and lower subsoil layers. Edwards and Lofty (1977) note
that in temperate climates, the deeper penetrating earthworms are Lumbricus terrestris and that Aporrectodea tuberculata frequent layers of the upper 40 cm. Taxonomic descriptions of soil pedons (Soil Survey Staff 1951, 1975)
normally notate the soil structure and biopore features,
but the volume of biopores is usually not described.

Structural features of the lower packed layer extend
into the upper subsoil and lower subsoil layers unless soil
texture changes, or a marked genetic horizon occurs. There
is usually an expression of larger soil structural units
with increasing depth through the subsoil; the trend may
start in the packed layer. Columnar and prismatic forms
are among the larger units of soil structure. Southard and
Buol (1988a) found subsoil ped size in Ultisols to increase
with depth and the grade of structure (differential between
cohesion within peds and adhesion between peds) to increase
with poor drainage and as silt content increased. The ped
size increase with depth was associated with decreasing
saturated hydraulic conductivity, but an increasing grade
of structure and a larger silt content did not influence
saturated hydraulic conductivity because inter-ped planes
were not sufficiently strong to contribute to saturated
flow (Southard and Buol 1988b).

Pedological Description of Soil Structure

Table 2.1 contains some typical descriptions of the soil structural properties according to the root zone layers of Figure 2.1. In this set are included soils from three soil orders--Ultisol, Mollisol, and Inceptisol. One soil, Norfolk, is known to be a difficult rooting soil because of the E horizon (Campbell, Reicosky, and Doty 1974), which coincides with the packed layer in Figure 2.1; the Cecil soil occasionally shows a tillage pan, most likely in the upper part of the packed layer. Table 2.1 shows that in general the tilled layer exhibits a granular/platy structure with angular fragments, while the packed layer shows some platy structure at the top grading into subangular blocky structure at the lower boundary. Some form of angular blocky, subangular blocky, or prismatic/columnar structure prevails in the subsoil layers. Note also an increase in grade number at these depths, especially in the poorly drained Typic Eutrochrept. The original profile descriptions of those used in Table 2.1 usually included layers below 80 cm. Generally these descriptions showed angular blocky, subangular blocky, and prismatic structure in the B horizon and a more massive structure in the C horizon. When interpreting soil hydraulic properties and rooting in these structural features, the texture, clay type, and porosity fraction made up of biopores must all be observed.

SOIL STRUCTURAL INFLUENCES ON ROOT ZONE
AND RHIZOSPHERE ENVIRONMENT

Soil environmental factors of water, heat, gaseous exchange, mechanical impedance, and available nutrients influence root development and function. Photosynthate supply from the shoot is also a major factor in root development and function, and again the system of shoot and root must be considered to determine when the supply of photosynthate is limiting rooting. Soil environments are determined by soil structure in arable soils; they are sensitive to interactions with meteorological conditions at the soil surface. Fluxes of water, heat, and gases for a single isotropic layer of soil will first be considered to demonstrate how water content, heat content, and gaseous concentration vary over time. Then a sketch of typical approaches for a multilayered soil (as in Figure 2.1) will demonstrate a system atic interpretation of structure influences on root-zone and rhizosphere environment.

Fluxes in a Single Homogeneous Layer

Within a homogeneous layer of the rooted zone shown in Figure 2.1, the flux of water, heat, or a diffusible gaseous component can be described with a linear flow law (Kirkham and Powers 1972; Hillel 1980), which has the general form

$$\upsilon = -Ki \qquad [1]$$

where υ is a flux per unit cross section (flux density) made up of a flux (Q) and a cross section (A), i is the gradient, and K is a transport coefficient.

For saturated water flow, equation [1] is the Darcy equation, for heat flow, it is the Fourier Law applied to heat flow by conduction, and for gaseous diffusion, it is called Ficks Law. Kirkham and Powers (1972) give details about sign convention and other adjustments needed, such as to account for porosity and tortuosity in a structured soil. A differential equation for the description of flow through a soil layer is developed from a linear flow law and a balance of flow with conservation of energy and mass. For saturated water flow it has the following differential form

$$\partial\theta/\partial t = \partial \ (K \ \partial h/\partial z)/\partial z \qquad [2]$$

where θ is water content (m^3/m^3), t is time (sec), K is the Darcy constant (m/sec), h is hydraulic head (m), and z is depth (m). In unsaturated soil K cannot be considered constant, but rather it is a function of the water content or water potential (Ψ).

The differential flow equation for heat flow by conduction is

$$\rho c \ (\partial T/\partial t) = \lambda(\partial^2 T/\partial z^2) \qquad [3]$$

where ρc is the volumetric heat capacity [J/m³ °K], T is temperature (°K), t is time (sec), z is depth (m), and λ is the thermal conductivity (Fourier constant, J/(m s °K). The form of equation [3] indicates that λ is not a function of z or t.

The common differential equation given for diffusion of a gas (diffusing substance) in soil is

$$\partial C/\partial t = D \, (\partial^2 C/\partial z^2) \tag{4}$$

where C is the concentration of a diffusing substance (kg/m^3), t is time (sec), z is depth (m), and D is the Ficks diffusion coefficient (m^2/sec). In this formulation D is a constant. Again there are refinements if D is not a constant, and addition of a sink term if the gas (oxygen, for instance) is consumed in the layer (Hillel 1980).

Details of these four equations are given to illustrate the important parameters needed to describe flow. For water flow these are K_{sat}, K_{unsat} — function of water content (θ) or potential (Ψ), and water content (θ) = f (Ψ); for heat flow they are ρc (heat capacity), λ (thermal conductivity), and temperature; for a diffusing substance they are D (diffusion coefficient) and concentration. The gradient or the divergence of the gradient in equations [2], [3], and [4] is determined by the initial and boundary conditions and the flow itself. It is important to note that heat and water flow in soil are linked because the heat content of soil affects soil water movement, and soil water movement transfers latent heat (see Philip and de Vries 1957 for a mechanistic approach).

Fluxes in a Layered Soil

In a layered soil such as the scheme of root-zone layers in Figure 2.1, it is difficult to apply the differential equations, and thus numerical simulations are used to describe flow. Van Genuchten (1978) developed a simulation of saturated/unsaturated flow of water in a layered soil using either a finite-difference or a finite-element method, and Wierenga and deWit (1970) simulated heat flow in a layered soil using a finite-difference approach. Gaseous movements in soil have been mathematically modeled for the whole soil (Hillel 1980; Campbell 1985), but gases such as oxygen and carbon dioxide generally move readily in interaggregate pores. There are therefore numerous mathematical simulations to describe gaseous movements to the root and into the smaller pores located inside aggregates.

A complete simulation of fluxes of heat and water in soil requires that these fluxes link with micrometeorological conditions at the soil surface, the upper boundary of the layered soil system. With respect to infiltration and

TABLE 2.1. Typical variations of soil structure in the various root-zone layers as observed pedologically.

| Taxonomic name/ state/ reference | Soil series | Structural designation of the various root zone layers* | | |
		Ti	Pa	SS
Typic Hapludult Georgia Bruce et al. 1983	Cecil	0-23 cm 1 m gr	23-33 cm 1 m sbk	33-100 cm 2 m sbk (argillic horizon)
Typic Haploxeroll Oregon Allmaras et al. 1982	Walla Walla	0-20 cm 1 gr to pl	20-40 cm 1 m sbk	40-100 cm 1 m sbk to 1 f sbk
Typic Paleudult North Carolina Southard and Buol 1988a	Norfolk	0-28 cm 1 f gr	28-46 cm 1 f gr	46-70 cm 1 m sbk (argillic horizon)

		0-15 cm	15-32 cm	32-78 cm
Typic Paleaquult North Carolina Southard and Buol 1988a	Rains	1 f sbk to 1 m gr	1 m sbk	1 m sbk (argillic horizon)
		0-20 cm	20-45 cm	45-75 cm
Typic Argiudoll Wisconsin Bouma, van Rooyen, and Hole 1975	Tama	angular fragments, gr	2 pl to 1 vf sbk	2 m pr to 3 f sbk
		0-20 cm	20-35 cm	35-70 cm
Typic Eutrochrept Wisconsin Bouma, van Rooyen, and Hole 1975	Oshkosh	angular fragments, gr	2 fpl to 3 vc pr	3 vc pr to 3 f bk

*A structural triad specifies (grade, class, type) where 1, 2, 3 is weak, moderate, strong; f, m, c, vc is fine, medium, coarse, very coarse; and pl, gr, sbk, b, pr, is platy, granular, subangular blocky, angular blocky, prismatic.

water movement into a soil through the surface soil, there are many simulations such as those by Moore (1981) and Brakensiek and Rawls (1983), each with somewhat different assumptions about the surface soil, especially with regard to transient features and surface seal formation. An accurate appraisal of heat flow requires that the energy and mass balance at the soil surface be linked to fluxes of heat and evaporative water loss. To accomplish this simulation there must be simultaneous flux of water and heat in the soil (McInnes et al. 1986; Bristow et al. 1986), each linked to transfers at the soil surface. Furthermore, the diurnal features of soil water content, soil temperature, and balances of energy and mass at the soil surface are necessarily determined.

Water Flux. Only under unusually large axle loads or when a deep cultivation is performed will the K_{sat}, K_{unsat} = f (θ), and Ψ = f (θ) of the Pa, SSu, and SSl layers be changed during tillage. Plant rooting in biopores may, however, change these properties. K_{sat} of the packed and subsoil layers can be estimated by a method of McKeague, Wang, and Topp (1982) using information about macropore area (as a percent of pores), the pedal or structural properties (Table 2.1) of a soil, and the texture. K_{sat} is often measured in undisturbed soil samples from these layers (Klute and Dirksen 1986). K_{unsat} = f (θ) and Ψ = f (θ) can both be determined from in situ internal drainage methods Green, Ahuja, and Chong 1986; Hillel 1980; Allmaras, Nelson, and Voorhees 1975; Arya, Farrell, and Blake 1975) for the Ψ range of about -2 to -30 kPa. The method must be supplemented by measurements of $\Psi(\theta)$ on undisturbed soil samples (Klute 1986) and by theoretical estimation of K_{unsat}, both in the range of Ψ = -30 kPa to -1.5 MPa to include the full range of θ in which roots are active. Perhaps the best guide about depth ranges in which these properties are unchanged is the soil morphological description including horizon depths. In fact, an estimation of K_{sat} using the McKeague, Wang, and Topp (1982) procedure may be the first and most direct method to guide the depth ranges for K_{unsat} = f(θ) and Ψ = f(θ). The parameters/functions of K_{sat}, K_{unsat} = f(θ), and Ψ = f(θ) of the tilled layer are subject to change each time there is tillage; furthermore, these hydraulic properties change with time as a result of meteorological inputs (wetting, drying, freezing, thawing) and biological activity. Perhaps the most systematic manner of estimating these hydraulic properties and the susceptibility to infiltration is that of Rawls, Brakensiek, and Soni (1983) and Brakensiek and Rawls (1983).

They used a combination of measured porosity and random roughness produced by different tillage combinations (All-maras et al. 1977) and estimates of $K_{unsat} = f(\theta)$ and $\Psi = f(\theta)$ made from a knowledge of soil constitutive properties of texture, carbon content, and effective saturation (see Horton, Allmaras, and Cruse 1989 for details). After the tilled layer has consolidated somewhat, the same approaches could be used as outlined above for the packed and subsoil layers.

A mechanistic approach toward understanding soil struc-ture influences on the water content aspects of the root zone is to make a simulation of infiltration and redistribu-tion (van Genuchten 1978) starting with a freshly tilled soil and proceeding until the potential root zone is wetted and/or significant runoff is produced. Such an approach in-tegrates the hydraulic properties of all root zone layers into a system, including those layers frequently modified by tillage as well as those not normally changed.

Figure 2.2 shows profiles of soil water content as in-fluenced by tillage of the tilled layer. Each of the four tillage treatments (the primary tillage is the identifier) gave different distributions of water content in the packed and subsoil layers on both measuring dates. The chisel treatment had more water storage than the other treatments; the till-plant, conventional, and no-till all had about the same storage but the depth distribution varied. These dis-tributions of water content were measured just before the maximal use of water by corn. Allmaras, Nelson, and Voorhees (1975) described the use of water by corn and soy-bean roots as they grew. This was done merely by placing a sink term $S(z,t)$ on the right side of equation [2]. Addi-tional water content profiles at different times of the year (Johnson, Lowery, and Daniel 1984) demonstrate that the water content in the root zone can increase and de-crease repeatedly in a growing season depending upon infil-tration and plant use.

Heat Flux. The thermal environment of the rooted zone (Figure 2.1) shows a different sensitivity to the tilled surface layer and lower layers than that noted above for the soil water environment. Because of the diurnal charac-ter of heat flux and continuously available heat sources and sinks at the surface, thermal properties of the soil surface and the tilled layer play a dominant role in heat-flux. The moisture environment of the rooted zone (Figure 2.1), however, plays a role in thermal environment--so much so that heat flux is difficult to describe without consid-

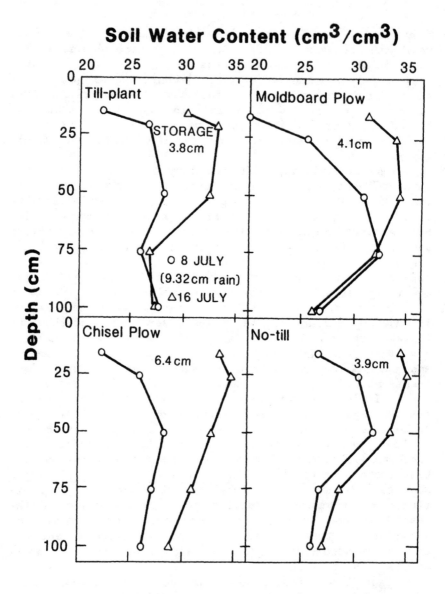

Figure 2.2. Typical changes in the soil water content of the soil profile (including tilled, packed, upper subsoil, and lower subsoil layers of Figure 2.1) as related to the management of tillage and residue cover.

Source: Johnson, Lowery, and Daniel 1984. Reprinted by permission.

ering coupled heat and water flow. The balance of energy at the soil surface determines the sinks/sources for thermal energy:

$$(1 - \alpha_s)R_s + R_1{}^{net} - R_n - H_{soil} + H_a + H_e \qquad [5]$$

where α_s is the reflection coefficient for incoming short-wave radiation; R_s is the incoming short-wave radiation; $R_1{}^{net}$ is the balance of long-wave radiation; and R_n is the net radiation available for soil heat flow (H_{soil}), heating the air (H_a), and evaporating water (H_e).

Soil management by tillage and the positioning of crop residues on the surface can influence all of the terms except R_s (Voorhees, Allmaras, and Johnson 1981). Solar zenith angle, soil color, soil roughness, mulches of crop residues, and soil water content all have a strong influence on α_s. A change in H_e must produce a change in H_{soil}, H_a, or both. A crop canopy and soil packing from traffic between rows will also modify the factors other than R_s in equation [5]. At a constant R_n and H_e, soil management can change the amplitude of the heat flux density at the soil surface in proportion to the change in thermal admittance $(\lambda \rho c \, \omega)^{\frac{1}{2}}$. The term ρc was defined in equation [3], ω is a constant descriptive of the period of oscillation (i.e., diurnal or annual), and λ as used here includes both conduction and heat transfer in water vapor transfer.

Equation [3] dealt with heat flux by conduction only. In a soil system with pores and water films, heat flux within a small depth increment (Δz) must be described as

$$\partial H(z,t) - \partial[-\lambda(\partial T/\partial z)]/\partial z - -(\Delta z)\rho c \, [\partial T/\partial t] \qquad [6]$$

If λ and c are unchanged with time and depth in this depth increment, equation [3] holds with a different λ to describe heat transfer by conduction and vapor movement. Both c and λ depend on such factors as soil porosity, water content, soil organic matter content, and mineral matter content and type (Allmaras et al. 1977; Voorhees, Allmaras, and Johnson 1981; Bristow et al. 1986); at excessively high porosity as with moldboard plowing and no secondary tillage, λ is underestimated by the method of de Vries and an enhancement factor may be obtained by correlations of random roughness and porosity (Allmaras et al. 1977). Because λ and c are derived directly from constitutive properties on a volume basis within each of the layers in the rooted zone (Figure 2.1), the parameters and functions required to compute heat flux are more readily derived than are those for water flux. As with water flux, the parameters (λ and c) of the packed and subsoil layers do not change each time that the tilled layer is tilled. As with parameters of water flux, thermal parameters in the tilled layer change as the tilled layer undergoes consolidation or other structural changes over time.

The coupled flow of heat and water is mechanistically described (Philip and de Vries 1957) as the interactive sum of liquid water flow due to a water potential gradient, heat flow due to a temperature gradient, and water vapor flow due to a water potential and a thermal gradient. Simulation modeling of coupled heat and water flow requires that each of these flow components is coupled throughout, including the interface between soil and air at the soil surface (McInnes et al. 1986; Bristow et al. 1986). Distinct advantages of this approach include a linkage between microclimate and the environments of heat and water in soil and an elaboration upon diurnal changes of thermal and water environment, especially within the tilled layer.

Figure 2.3 illustrates some typical diurnal variations of soil temperature as related to tillage and surface mulch variations. These measurements demonstrate tillage influences on soil thermal properties which in turn influence the diurnal amplitude and the phase lag of the sinusoidal trace. Soil temperature maximums lag as depth increases. Maximum insolation occurs at about one-half hour past noon. There is also some increase in T_{10} over T_{100} during the day, showing a sink for heat into H_a. The observed amplitude of temperature at the base of the tilled layer shows soil heat flux into the subsoil (packed and subsoil) layers; energy balance components (Table 2.2) also show H_{soil} into the subsoil. Net radiation (R_n) in Table 1.2 reflects the influence of tillage and surface mulch on α_s mainly and somewhat on R_1^{net} in equation [5]. Even on a bare surface, the sum of $H_a + H_e$ ranged from 71 to 83 percent of R_n (Table 2.2). Under a crop canopy this value is more likely to be 90 percent or greater.

In the same manner as with a diurnal soil-temperature wave (Figure 2.3), the amplitude of the annual soil temperature wave at 120 cm is about 7°C and the phase lag at this

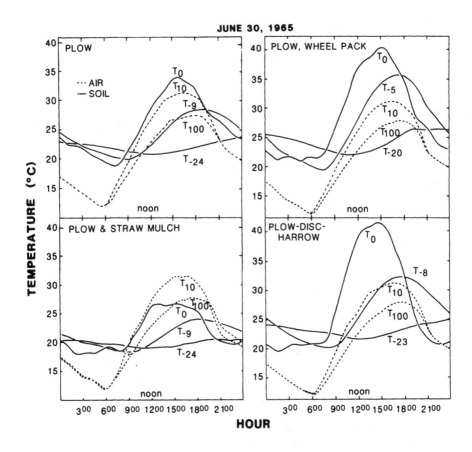

Figure 2.3. Typical influence of tillage and straw mulch on the diurnal variation of soil temperature in the tilled layer. (The depth of temperature record is coded with positive values above the surface and negative values below the surface.)

Source: Allmaras et al. 1977. Reprinted by permission.

depth is about one month in a Mollisol at 44°N latitude (Allmaras, Nelson, and Voorhees 1975). The maximum and minimum soil temperatures do not vary more than 2°C from year to year. At this depth the soil temperature is increasing during the period of rooting development of spring seeded crops, and perched water tables may exist after the long dormant period for water recharge of the root zone. Advance of the rooting front by corn and soybeans in such a soil environment may be controlled by a

TABLE 2.2. Components of energy balance during the positive net radiation period for tillage treatments shown in Figure 2.3[*].

Tillage treatment	R_n	H_{soil}		H_a	H_e[**]
		Ti layer	Subsoil		
		-------------	MJ/m^2	-------------	
Moldboard plow	19.7	2.6	0.6	9.6	7.1(2.8)
Plow and pack	18.2	4.5	0.7	8.0	5.0(2.0)
Plow, disk, harrow	18.9	4.0	0.6	8.0	6.3(2.5)
Plow and mulch	16.3	2.1	0.4	8.0	5.9(2.3)

[*]Symbols explained in equation [5].

[**]Values in parentheses are mm of water.

minimum soil temperature and/or maximum water content consistent with aeration (Allmaras, Nelson, and Voorhees 1975). Kuchenbuch and Barber (1988) found that corn root distribution at silking stage was significantly different among nine years at the same site in Indiana. Growing degree days (air temperature) within three weeks after planting explained rooting density below 30 cm, and precipitation during a three-week period preceding silking explained rooting density above 30 cm. In this way year-to-year variation of growing degree days (a surface temperature function) influenced rooting in the subsoil even though there was negligibly small year-to-year variation of soil temperature at 80 cm.

<u>Gaseous Flux</u>. Respiration and growth of roots and aerobic microorganisms require that oxygen diffuse from the soil surface through the soil pores to the microsites of root and microbial activity. Carbon dioxide must move in the reverse direction away from these sites where it is being produced during respiration. Diffusion coefficients for carbon dioxide and oxygen in air are roughly 10^4 greater than for these same gases in water films around roots and soil microbes. The soil moisture regime in the root zone model (Figure 2.1) is a primary influence on the

availability of oxygen for respiration and the facility to diffuse carbon dioxide away from microsites with respiring organisms. When a soil is very wet and pores are nearly water-filled, the oxygen supply is used more rapidly than it can be replaced by convection and diffusion. When there is air-filled porosity above a limiting value of soil water content, oxygen exchange to the microsite is not limiting except in the vicinity of respiring organisms at the micro-site. Soil temperature mainly determines the biological demand for oxygen and the amount of carbon dioxide to dif-fuse away from microsites. (Typical biological reaction rates double for each 10°C increase. Although soil temp-erature may influence diffusion coefficients for these gases in water films, that influence plays a minor role compared to the influence on biological reaction rate.)

Because of the relatively faster exchange of oxygen and carbon dioxide in the interaggregate pore system than into aggregates themselves or though water films around roots and microbes, the bulk-soil flow equations such as in equation [4] have limited application (Campbell 1985). Rather, steady-state flow with negligible storage of the gas in a finite volume can be approximated using the differential form of the linear flow law of equation [1]. The flux (Q, g/s) is then (Campbell 1985)

$$Q = K (C_2 - C_1) \qquad [7]$$

where C is concentration (kg/m^3), and K is a conductance that is the reciprocal of the resistance.

When diffusion is planar, K equals $D/(X_2-X_1)$; when dif-fusion is spherical, as in a cluster of microorganisms or in an aggregate, K equals $4\pi D \; r_1 \; r_2/(r_2-r_1)$; and when diffusion is cylindrical, as in a root, K equals $2\pi D/\ln(r_2/r_1)$. D is the diffusion coefficient with units m^2/s, while X and r have units of meters. The units of K differ depending on geometry: for planar diffusion, K has units of m/s and flux per unit area is constant; for spherical diffusion, K has units m^3/s and flux is per unit sphere; and for cylindrical diffusion, K has units m^2/s and flux is per unit length. Each of these conduct-ances can be combined with equation [7] to describe gaseous transport in the profile as a whole (planar geometry), into an aggregate or a cluster of microorganisms (spherical geometry), or into a root (cylindrical geometry). A trial application of the planar diffusion of oxygen in a soil profile showed that as air-filled porosity dropped below 0.2 m^3/m^3, a significant decrease in oxygen in the soil matrix was expected (Campbell 1985).

Compacted soil frequently occurs between rows where there is traffic and also in the lower part of the tilled layer and upper part of the packed layer (Figure 2.1). Besides a reduced total porosity, a compacted soil also has reduced internal drainage, lower air-filled porosity, and a reduced movement of oxygen and carbon dioxide by mass flow (convection) and diffusion (Smucker and Erickson 1989). Schumacher and Smucker (1981) relate that increased imped-ance to root growth increases the oxygen uptake by roots (specific demand for oxygen). Poiseuille's Law applied to capillary flow shows that gas flow through a single pore with a radius of 1 mm is equal to the flow through 10^4 capillaries each with a radius of 0.1 mm and a total cross-sectional area 100 times greater (Glinski and Stepniewski 1985). Soil compaction may produce such a change in pore geometry.

An anaerobic environment in soil microsites may develop within spherical aggregates in a structured and moist soil with interaggregate spaces, within a spherical cluster of microorganisms surrounded with a water film, or within a root contained within a water film. Several methods may be used to indicate anaerobic microenvi-ronments, including (1) oxygen diffusion to a platinum microelectrode inserted into soil [critical values less than 33 $\mu g/(m^2\ s)$], (2) air-filled porosity less than 0.1 m^3/m^3, or (3) D/D_0 approaching zero where D is the diffusion coefficient of oxygen measured in soil and D_0 is the diffusion coefficient of oxygen in air. The ratio D/D_0 is generally not dependent upon the nature of the diffusing gas. Smucker and Erickson (1989) show that D/D_0 is a more sensitive index of aeration at higher soil bulk density and that oxygen diffusion to a microelectrode is a more sensitive index at low bulk density. Without inter-vention to obtain undisturbed samples, the possibility of anaerobic conditions can be estimated by examination of the water content of the root-zone scheme in Figure 2.1 based on the water flux. Whenever total pore space is less than 0.1 m^3/m^3 air-filled, at least a brief anaerobic en-vironment can be expected. For any given soil layer, this air-filled pore space (ξ, m^3/m^3) might be estimated

$$\xi = 1 - (\rho_\beta/\rho_\alpha) - P_w \qquad [8]$$

where ρ_β is soil bulk density (Mg/m^3), ρ_α is the density of soil solids (Mg/m^3), and P_w is the water content (m^3/m^3).

An especially useful measure of the continuity for oxygen diffusion is a measure of D/D_0 as a function of ξ at various depths in a soil profile. This measurement requires the extraction of undisturbed soil cores (Douglas et al. 1986).

Figure 2.4 typifies the influences of wetting and drying and bulk density upon the diffusion of a gas (i.e. hydrogen) in an aggregated medium (Currie 1984). For each dry bulk density, the bed of soil aggregates was either wetted (reducing air content) or dried (increasing air content) beginning at the intersection of the two curves. In one curve D/D_0 is nearly a linear function of ξ, and in the other curve, D/D_0 was a power function of ξ. As the bed of aggregates was wetted, beginning at the intersection of curves, the interaggregate pores were being filled with water and thus reducing ξ. As the bed of aggregates were dried, beginning at the intersection of curves, water was being removed from the interaggregate pores and ξ was increasing. When wheel traffic is produced between rows as in Figure 2.1, the bed of aggregates becomes more compacted. The two curves for one dry bulk density thus become less distinct because interaggregate pore space is being reduced.

There are many chemical and biological reactions in soil that are sensitive to oxygen concentration; these reactions affect the oxidation state of elements in the soil, produce compounds toxic to roots, and control availability of plant nutrients (Glinski and Stepniewski 1985). Many of these reactions can be linked to measurements of redox potential.

Mechanical Strength of Soil in the Root Zone

A soil structure mechanically resistant to root development can occur at nearly all depths shown in Figure 2.1. Hamblin (1985) suggests that mechanical resistance to root development is ubiquitous. Soil mechanically resistant to roots may be located in the tilled layer and upper part of the packed layer as a result of compaction from vehicular traffic or shear near maximum penetration of tillage tools. In the packed and subsoil layers soil may be mechanically strong due to soil genetic features in argillic, illuvial, fragipan, and duripan horizons. Constraint upon the development of root systems in these layers (Figure

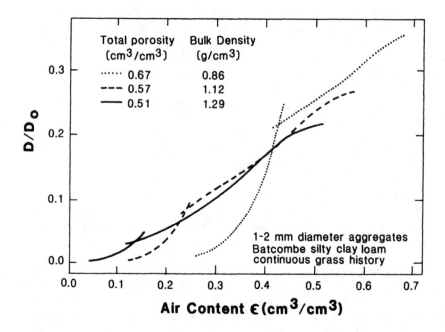

Figure 2.4. D/D_o for H_2 as a function of ξ that was varied in a bed of aggregates by wetting or drying and also by compaction.

<u>Source</u>: Currie 1984. Reprinted by permission.

2.1) depends upon the water flux and water content (see the earlier section, "Water Flux") as well as the degree of structure manifested by cracks, planar voids, biopores, packing voids, and pedal properties (Table 2.1). Such soil structure features limit somewhat the fundamental notion that roots must exert a pressure to open a channel for elongation; rather, roots may be elongating and branching throughout a continuous system of weak interfaces produced by soil structure features. Root elongation by generation of cylindrical root pressures in a continuum, in which the interparticle distances are only a fraction of the diameter of the cavity produced by rooting, has been studied in detail (Greacen, Barley, and Farrell 1968). Root growth following a path of weak interfaces will be discussed in a later section.

Compaction influences soil structure more through changes in pore geometry than through changes in bulk den-

sity (Gupta, Sharma, and DeFranchi 1989). Some soil properties that are influenced by compaction include water retention characteristic, infiltration, sorptivity, and gas
diffusion. Soil water content, clay type and amount, bulk
density, and the roughness of soil particles all influence
soil strength, as expressed in the shear strength

$$\tau = C + \sigma_n \tan \phi \qquad [9]$$

(τ, kg m^{-2}) where C depends on soil cohesive factors,
σ_n is the normal load acting on the plane of shear, and ϕ
is the angle of internal friction. C depends upon surface
area of clay (clay content and clay type), soil water content, and bulk density of the soil; ϕ relates to the force
needed to move rough particles past each other and is related to soil water content, packing arrangement, and
angularity of particles. C is also increased by cemen-
tation of soil materials, such as silica or iron oxide,
during drying. Shear failure is a common mode of soil
strain occurring under tractor tires and when roots must
enlarge a cavity to accommodate their elongation.

An application of a uniaxial stress (σ_1) to an
assembly of soil aggregates constrained laterally (σ_3
greater than σ_1) in a cylinder is a method of measuring the
compression index, which is the slope of a linear relation
between the void ratio and ln σ_1 (Gupta and Allmaras
1987). Soil water content merely changes the intercept of
this linear relation. Compression index is therefore a
soil characteristic indicative of the susceptibility to
compaction.

The result of a series of uniaxial stresses (σ_a) on
soil aggregate assemblies with different initial water
contents is illustrated in Figure 2.5 for a Nicollet clay
loam with a compression index of 0.59. Each curve, where σ_a
equals applied stress, shows the soil porosity remaining
after the stress was relieved. For each water content a
vertical traverse above the "saturation" line corresponds
to a decreasing bulk density and increasing air-filled
porosity. A traverse to lower water contents at a constant
porosity replaces water and air at a constant bulk density.
All ordinate values between the "saturation" line and "10
percent air-filled porosity" identify poor aeration and
potential anoxia; all ordinate positions below the "critical penetration resistance" and above the "saturation"
lines are in a region of water content and porosity wherein
soil mechanical strength is sufficient to stop root growth
as determined by a critical pressure of 2 MPa on a cone pen

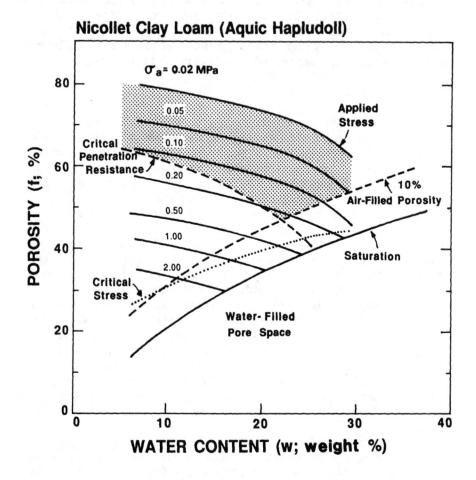

Figure 2.5. Porosity components as a function of water content (W; percent by weight) in a Nicollet clay loam (31 percent clay and 2.2 percent organic carbon) after different uniaxial stresses were imposed. Superimposed are criteria of 10 percent air-filled porosity, critical penetration resistance, and stress critical for shearing aggregates.

Source: Gupta and Larson 1982. Reproduced from "Predicting tillage effects on soil physical properties and processes". ASA Special Publication 44, 1982, p. 151-178 by permission of American Society of Agronomy, Inc. and Soil Science Society of America, Inc.

etrometer; and all ordinate positions below the "critical stress" and above the "saturation" lines are in a region of water content and porosity in which the stress was large enough to shear aggregates. More details of interpretation are given by Allmaras, Kraft, and Miller (1988), Gupta and Allmaras (1987), and Gupta and Larson (1982).

Cone penetrometers (Bradford 1986) are often used to estimate soil strength as a factor for mechanical resistance to root growth. In Figure 2.5 a pressure of 2 MPa on a cone penetrometer with a 30° included cone angle was designated as a "critical penetration resistance." A cone penetrometer pressure of 2 MPa is commonly accepted as that corresponding to cessation of root growth, while measured actual root pressure may range from 0.2 to 0.5 MPa (Richards and Greacen 1986). The critical cone penetrometer pressure ranges from 1 to 4 MPa depending on pore water pressure (related to soil water potential), mechanical composition (sand, silt, and clay contents), and also plant species (Hamblin 1985). Another factor is macroporosity, which will be discussed later. The range of 1 to 4 MPa critical pressure can also be influenced by associated aeration, as demonstrated by Voorhees, Farrell, and Larson (1975).

In a root zone as depicted in Figure 2.1, cone penetrometers have been used to map out zones of high strength. Voorhees, Nelson, and Randall (1986) located layers of greater soil strength as related to compaction in soil layers, especially that corresponding to the packed layer; Busscher and Sojka (1987) developed iso-pressure diagrams from a multitude of systematically spaced probe penetrations to show soil strength responses to row orientation, traffic, tillage tool, and E horizon transformations. he E horizon is a coarse-textured horizon susceptible to high strength development because of its particle size distribution and low content of organic matter. Soil moisture profiles are often changed by the same factors that change cone penetrometer pressures; consequently concomitant soil water content measurements are necessary to interpret soil strength changes not accountable by soil moisture changes.

ROOT DEVELOPMENT IN THE SPATIALLY VARIABLE
ROOT ENVIRONMENT

Recent reviews consider the general form of the root system and its development (Hamblin 1985; Rendig and Taylor

1989, Chapter 2; Taylor and Klepper 1978). At germination the radicle (primary or seminal root) encounters the tilled layer. Soil environment in this layer may require that the root enlarge a cavity for elongation, or the root may grow between interfaces produced by soil structure. Root elongation, branching, diameter, and orientation would reflect this growth pattern. As the root system progresses through the soil layers described in Figure 2.1, various stresses may be encountered that may reduce growth of the whole root system. In addition, compensatory root growth occurs due to heterogeneous soil conditions. Root elongation and branching may be increased in an environment that is more favorable while growth is simultaneously reduced in an unfavorable environment (Hamblin 1985). These variable soil environments and compensatory growth occur not only from one layer to the next but also within a given layer.

Root Response to Horizontal Heterogeneity in the Tilled Layer

Low soil temperature dramatically reduces root dry matter accumulation (Voorhees, Allmaras, and Johnson 1981; Walker 1969), elongation rate (Logsdon, Reneau, and Parker 1987), and rate of branching. The maturation zone (consisting of suberized endodermis, branch root formation, and secondary tissue) occurs closer to the root apex at low temperatures. Root orientation also becomes more horizontal at low temperatures. As temperature increases, root angles from the vertical decrease (Sheppard and Miller 1977).

The tilled layer in a corn field is subject to more extremes in temperature than deeper layers. Soil temperature is influenced by soil color, surface random roughness, surface residue, and soil moisture content. Allmaras et al. (1987) showed that in the top 10 cm, corn roots did not spread into the interrow if the interrow had a low bulk density (moderate temperature) or if the row was mulched (cooler and wetter soil) (Figure 2.6). Roots did spread into the interrow between 10 and 20 cm in the low bulk density treatment, but roots did not spread very far into the interrow if the row was mulched. Chaudhary and Prihar (1974) observed more lateral spread of roots in a soil mulched over all because the lower soil temperatures there were desirable for root growth. Drew and Saker (1980) noticed similar trends for a mulched and no-till treatment compared with an unmulched and moldboard plow treatment.

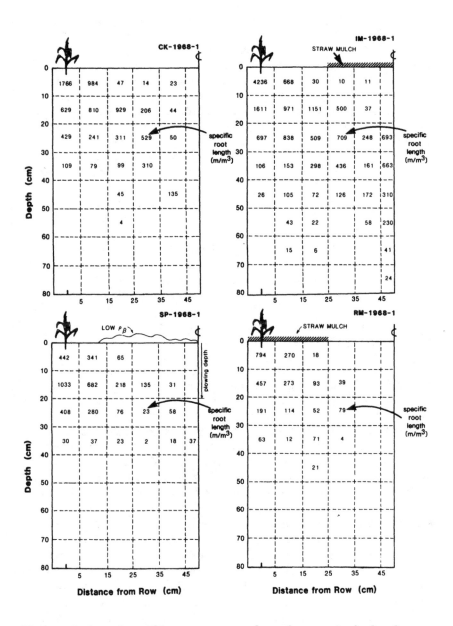

Figure 2.6. Specific corn root length at sixth leaf stage for four selected interrow variants in 1968.

Source: Allmaras et al. 1987.

The different root growth patterns where the mulch was placed in strips is a manifestation of compensatory growth.

High soil strength reduces root elongation. In rigid growth media roots cannot penetrate pores smaller than the diameter of the root (Russell and Goss 1974), but rigid media would be rare in a tilled layer. Impeded roots have a distorted morphology manifested by a slowed elongation and thickened root mass behind the meristem (Russell and Goss 1974). The dry matter of impeded roots (e.g. field beans, sugar beets, and spring barley) often does not decrease even though total length decreases (Brereton, McGowen, and Dawkins 1986).

Penetrometers are often used to estimate soil strength constraints on root growth, with critical pressures ranging from 1 to 5 MPa sufficient to prevent root growth (Greacen, Barley, and Farrell 1968; Bradford 1986). Russell and Goss (1974) considered cone pressure measured by penetrometers too high, because in a confining system with flexible walls filled with glass beads (small pores), root growth was reduced by as little as 20 kPa applied confining pressure. Richards and Greacen (1986) have since shown that the radial confining pressure is not equal to root pressure (which for a rubber root-analogue was 500 kPa).

The entire discussion of soil strength is only applicable for a homogeneous deformable matrix and is not directly related to soils with macrostructure. Ehlers et al. (1983) observed different critical penetrometer resistance in the uppermost [Ti] layer for tilled and no-till soils. The critical penetrometer resistance for no-till was higher, (4.6 to 5.1 MPa, compared with 3.6 MPa for tilled soil) because continuous wormholes in the no-till treatment provided low-resistance pathways for root growth. Roots can grow into low strength aggregates (Misra, Dexter, and Alston 1986), but only fine roots are able to grow into intrapedal pores (Edwards, Fehrenbacher, and Vavra 1964). Roots are less able to penetrate aggregates as aggregate size and/or strength increases (Misra, Dexter, and Alston 1986; Misra, Alston, and Dexter 1988). As aggregate size increases, the interior of aggregates may become anaerobic (Currie 1966); thus aeration may also be a factor reducing root penetration into aggregates. Large-diameter roots can displace fine aggregates from their paths (Logsdon, Parker, and Reneau 1987; Misra, Alston, and Dexter 1988); otherwise, roots are deflected around aggregates.

The soil strength of a tilled layer is often not uniform, and rooting is also variable in the tilled layer. Examples of soil with high bulk density in the tilled layer include a nearly homogeneous layer formed by recent traffic over the interrow (Tardieu 1988), and clods of high bulk density present from previous compaction before tillage (Tardieu 1988) or from compaction when the soil was too wet during tillage. Roots were often deflected around these high bulk density clods. Corn roots were reduced in the interrow of the uppermost layer for no-till compared with a moldboard plow treatment, more so under wheel traffic (Voorhees 1989). Soybean roots in the interrow of the uppermost layer were most dense for a no-till and non wheel-tracked treatment, indicating that the optimum degree of compaction for soybean roots is higher than for corn roots. Ellis et al. (1977) observed a reduced spread of barley roots into interrows under no-till because of greater impedance in the interrow. Interrow compaction treatments forced corn roots to concentrate under the plant and reduced spread into the interrow (Chaudhary and Prihar 1974).

Impedance stress is highly interactive with aeration stress. An excellent review of impedance and aeration is given by Cannell (1977). Impedance causes increased dry matter per length of root, which reduces internal aeration (Schumacher and Smucker 1981). Root contact with the liquid plus solid phases increases as a root encounters mechanical resistance, which reduces the area of root surface in contact with the gaseous phase beyond that implied from the volumetric components of air, solid, and water in the soil (Hamblin 1985). Compaction also encourages root disease (Allmaras, Kraft, and Miller 1988), both because the stressed root is weakened, and the pathogen exhibits greater tolerance toward anoxia. Cell wall plasticity of root cells is reduced as oxygen tension drops below 14 kPa (McCoy 1987).

Air-filled pore space decreases as soil saturation with water increases. As a rough guide, air-filled porosity must be greater than or equal to 10 percent ($D/D_o - 0$) to retain aeration (Grable and Siemer 1968). In compact soils air-filled porosities even greater than 10 percent can limit root growth. Asady, Smucker, and Adams (1985) observed a linear decrease in root penetration of compact layers as air-filled porosity decreased from 30 to almost 0 percent. Root elongation is more sensitive to ODR (oxygen diffusion rate) than oxygen concentration and can be adversely influenced by diffusible metabolites of anaerobic respiration. In the field environment anaerobic conditions and ODR are transitory (several days at most) but devastating. Root growth is restricted by ODR less than 58 $\mu g/(m^2$ s), with no elongation below 33 $\mu g/(m^2$ s) (Erickson and Van

Doren 1961). Oxygen diffusion rapidly dropped below 33 $\mu g/(m^2\ s)$ within 24 hrs after a 7 cm rain on a compact treatment (Erickson, 1982). Recovery to levels higher than 58 $\mu g/(m^2\ s)$ took between six and eight days for the compacted treatment, but only two days for a deep chisel treatment.

Aeration patterns are not uniform. Anaerobic sites may develop inside aggregates (due to slow diffusion) or where available carbon sources occur due to crop residue burial or the presence of dead roots and exudates. Compensatory root growth is common because of the variable aeration patterns (Schumacher and Smucker 1984).

Soil water content and potential may change frequently in any part of the root zone, and they have a major influence on root development and death. Water deficiency may increase the root/shoot ratio because of a greater reduction in shoot growth than root growth. Increases in soil strength as the soil dries have a major impact on root penetration. Roots continue to grow where soil water contents correspond to soil water potential ranging from -1 kPa to 1 MPa, and roots can regenerate after soil is rewetted. Emergence of root branches from inside the stele is a natural feature of this regeneration. This regenerative feature is important for the tilled layer, which is subject to the greatest extremes in soil moisture. The root extension rate is fairly constant even to water potentials of -500 kPa or lower (depending on the crop and soil), but root length decreases due to shedding of roots starting at about -100 kPa. Root weight and volume decrease steadily as the water potential falls from -5 kPa to -1000 kPa because the relative water content of the root is reduced to less than one-half (Taylor 1983; Russell 1977). As the soil is dried beyond -500 kPa water potential, water is mainly used for local root metabolism.

Nodal root growth of a monocotyledon continues into and through the tilled layer (and below) even as the plant matures. In contrast, growth of the root axis of a dicotyledon generally progresses deeper into the soil as the plant ages, and growth of large roots does not continue in soil near the surface.

Root Response to Vertical Heterogeneity in the Root Zone

Root growth into the packed layer of Figure 2.1 is reduced when no macropores are present. Increased overburden pressure results in a more confined system, so greater root pressures are needed to penetrate the soil (Whiteley and Dexter 1984). Thus roots are better able to penetrate the packed layer if the tilled layer has fine aggregates (Dexter 1986). Root growth through a smeared layer is reduced (Prebble 1970), especially as the soil dries, since soil strength increases as the soil dries. Desiccation cracks may form as the soil dries, which allows an alternate pathway for root growth (Tardieu 1988). Also, a vertical root may generate enough pressure to produce vertical fractures in a horizontally dense soil layer, especially if unconfined (Barley, Farrell, and Greacen 1965). Platy structured soil formed as a result of compression often does not have vertical planes of weakness for root growth. Whiteley and Dexter (1983) have shown that elongation of a pea root radicle into horizontal planes is much less than elongation into vertical planes. No macropores may be present in a sandy soil or a soil with nonexpanding clays such as in a dense E horizon. Slit tillage through the dense E horizon facilitated root penetration below this layer if the subsoil was not adversely acid (Elkins and Hendrick 1983).

Roots are often spared the difficulty of penetrating the packed layer below the tilled layer by being deflected until a crack or pore is encountered (Dexter 1986). Roots are able to grow in interpedal voids along ped faces (Bouma and Dekkar 1978; Drew and Saker 1980). In a dense, massive soil, roots are restricted to desiccation cracks (Figure 2.7).

Biopores formed by invertebrates and/or old root channels provide another pathway for root growth (Drew and Saker 1980). Biopores often extend more deeply into the soil than cracks (Van Stiphout et al. 1987). Wang, Hesketh, and Woolley (1986) observed that soybean roots that had entered burrows grew down to the end of the burrows. Taproots and first-order laterals died if they had failed to enter burrows before they reached 45 to 55 cm depth. The burrows appeared to be old earthworm channels that had been reused by roots year after year. Ehlers et al. (1983) observed oat roots where growth in the B horizon was restricted to earthworm channels.

Root growth proceeds into the subsoil layers as roots and evaporation dry out surface soil layers. As the soil dries, the temperature and aeration improve, cracks appear, and root growth becomes more vertical. A shadow effect may have occurred beneath dense surface layers (Tardieu 1988) due to root growth in continuous vertical biopores or

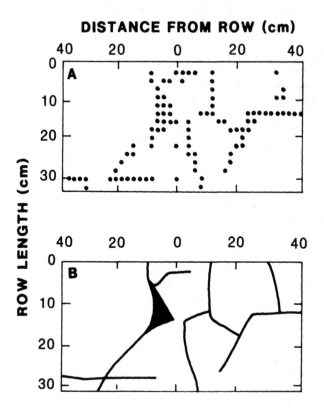

Figure 2.7. Mapping on a horizontal plane of a compact treatment at 17 cm depth A. Each point represents presence of roots in a 2 x 2 cm grid; B. Soil structural map.

Source: Tardieu and Manichon 1986. Reprinted by permission.

cracks (Figure 2.8). Pedal structure is often better in the subsoil layers and will provide low-resistance pathways for root growth around peds. Compensatory root growth at lower soil depths often occurs when roots have dried out surface soil layers (Figure 2.9). Such a growth pattern alters the often-assumed exponential decline of roots with depth (Gerwitz and Page 1974) that occurs under ample moisture conditions, as in frequent irrigation. Compensatory corn and soybean root growth due to wheel traffic was manifested by reduced root length in the surface layers asso-

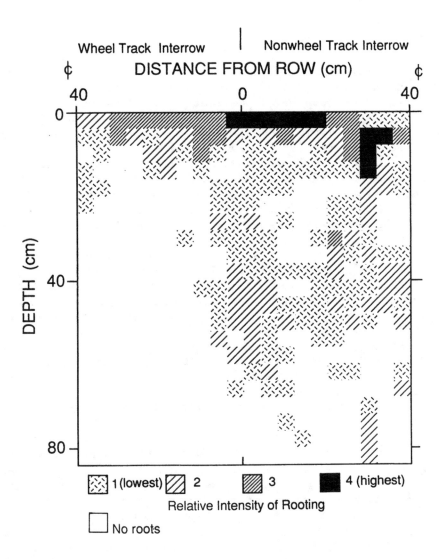

Figure 2.8. Vertical mapping of a compact treatment show-ing a "shadow zone" caused by a structural obstacle within the plowed layer.

Source: Altered from Tardieu and Manichon 1987.

ciated with increased root length in the subsoil (Voorhees 1989).

Root Growth Models

Simulation models have been proposed and used to study root system responses to soil environment. Root growth models have been developed that describe root initiation, elongation, and branching of roots in an unstressed, homogeneous medium (Hackett and Rose 1972; Lungley 1973; Narda and Curry 1981). Hillel and Talpez (1976) included root death. Models of root elongation and branching in homogeneous soil cannot be used in situ because soil structural factors in three dimensions are not being used to determine root responses to a path of least resistance.

Other models base root development on a balance of photosynthate supply and various resistances to root growth. For example, RUTGRO, the root growth subroutine for GOSSYM (Whisler, Lambert, and Landivar 1982) describes two-dimensional cotton root growth from cell to cell in response to soil physical conditions (impedance, oxygen level, moisture content) in that cell. ROOTS subroutine for NTRM (Shaffer and Larson 1987) allows for compensation in one-dimensional corn root growth. Interactions of roots and soil structure are not considered in these two models; moreover, input constants and model development were often speculative due to lack of supporting data.

Some models have included the influence of soil structure and macropores. Whiteley and Dexter (1984) modeled the elongation of roots in aggregated soil, accounting for roots that entered aggregates, were deflected around aggregates, or deflected aggregates from their paths. Jakobsen and Dexter (1987, 1988) wrote simulation models of wheat root growth, water uptake, and grain yield, incorporating the influence of subsoil structure, cracks, and biopores on root growth. Models predicting root response to tillage should include factors of root orientation due to temperature and growth in macropores, root response to physical and chemical stresses produced by tillage, and compensatory root growth under heterogeneous soil conditions.

The root zone is a heterogenous soil environment created by both natural causes and cultivation. Traffic and tillage influence both vertical and horizontal soil variations. Vertical variation may also be due to genetic horizons, and horizontal variation may be due to architecture of the plant canopy. Interactions among heat and

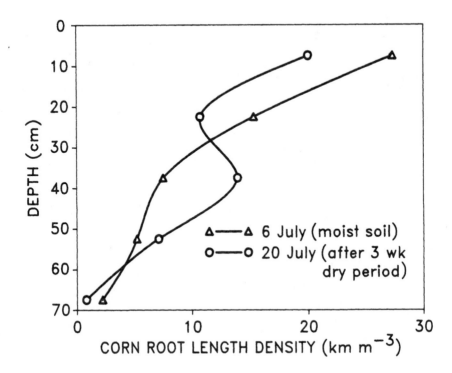

Figure 2.9. Corn root length density with depth at two dates showing the development of compensatory root growth deeper in the soil when the surface soil layers have dried.

water flow, gaseous flux, and soil strength determine the local soil environment affecting root growth. Roots themselves impose an environmental change. Roots respond not only to their local environment, but also as a root system. Compensatory root growth occurs when roots proliferate in a more favorable soil environment because part of the root system is restricted in a less favorable environment. Root response to isolated factors in homogeneous soil may not represent root response in arable agriculture because of these soil physical interactions, and because of compensatory growth. Future research needs in understanding root response to soil structure should include better knowledge of soil structure responses to management practices, as well as the root growth pattern and interaction with the three-dimensional soil structure features in the field. Simulation models of root growth compared with actual root responses may indicate where further study is necessary. Beyond this, simulation models may project influences of the root system on water and nutrient use, leaching of agricultural chemicals, and utilization of photosynthate.

REFERENCES

Allmaras, R. R., P. M. Burford, J. L. Pikul, Jr., and D. E. Wilkins. 1988a. Tillage tool effects on incorporated wheat residues. Proceedings of the eleventh conference of the international soil tillage research organization, vol. 2:445-450. Edinburgh:ISTRO.

Allmaras, R. R., S. C. Gupta, J. B. Swan, and W. W. Nelson. 1987. Hydro-thermal effects on corn root development in a Mollisol. In Transactions of the thirteenth congress of the international society of soil science, Symposia Papers, vol. 5:3-19.

Allmaras, R. R., E. A. Hallauer, W. W. Nelson, and S. E. Evans. 1977. Surface energy balance and soil-thermal property modifications by tillage-induced soil structure. Minnesota Agricultural Experiment Station Bulletin No. 306. St. Paul. 44 p.

Allmaras, R. R., J. M. Kraft, and D. E. Miller. 1988. Effects of soil compaction and incorporated residue on root health. Annual Review of Phytopathology 26:219-243.

Allmaras, R. R., W. W. Nelson, and W. B. Voorhees. 1975. Soybean and corn rooting in southwestern Minnesota. Soil Science Society of America Proceedings 39:764-777.

Allmaras, R. R., J. L. Pikul, Jr., J. M. Kraft, and D. E. Wilkins. 1988b. A method for measuring incorporated crop residues and associated soil properties. Soil Science Society of America Journal 52:1128-1133.

Allmaras, R. R., K. Ward, C. L. Douglas, Jr., and L. G. Ekin. 1982. Long-term cultivation effects on hydraulic properties of a Walla Walla silt loam. Soil and Tillage Research 2:265-279.

Arya, L. M., D. A. Farrell, and G. R. Blake. 1975. A field study of soil water depletion patterns in the presence of growing soybean roots. Soil Science Society of America Proceedings 39:424-430.

Asady, G. H., A. J. M. Smucker, and M. W. Adams. 1985. Seedling test for the quantitative measurement of root tolerances to compacted soil. Crop Science 25:802-806.

Barley, K. P., D. A. Farrell, and E. L. Greacen. 1965. The influence of soil strength on the penetration of a loam by plant roots. Australian Journal of Soil Research 3:69-79.

Bouma, J., and J. L. Anderson. 1973. Relationships between soil structure characteristics and hydraulic conductivity. In Field soil water regime, ed. R. R. Bruce, K. W. Flack, and H. M. Taylor. 77-105. Special Publication no. 5. Madison, WI: Soil Science Society of America.

Bouma, J., and L. W. Dekkar. 1978. A case study on infiltration into dry clay soil. Geoderma 20:27-40.

Bouma, J., D. J. van Rooyen, and F. D. Hole. 1975. Estimation of comparative water transmission into two pairs of adjacent virgin and cultivated pedons in Wisconsin. Geoderma 13:73-78.

Bradford, J. M. 1986. Penetrability. In Methods of soil analysis: physical and mineralogical methods, 2nd ed., pt. 2. ed. A. Klute, 463-478. Madison, WI: American Society of Agronomy.

Brakensiek, D. L., and W. J. Rawls. 1983. Agricultural management effects on soil water processes. Transactions of American Society of Agricultural Engineers 26:1753-1757.

Brereton, J. C., M. McGowen, and T. C. K. Dawkins. 1986. The relative sensitivity of spring barley, spring field beans, and sugar beet crops to soil compaction. Field Crops Research 13:223-237.

Bristow, K. L., G. S. Campbell, R. I. Papendick, and L. F. Elliott. 1986. Simulation of heat and moisture transfer through a surface-residue system. Agricultural and Forest Meteorology 36:193-214.

Bruce, R. R., J. H. Dane, V. L. Quisenberry, N. L. Powell, and A. W. Thomas. 1983. Physical characteristics of soils in southern region: Cecil. Southern Cooperative Series Bulletin no. 267. Georgia Agricultural Experiment Station, Athens.

Buntley, G. J., and R. I. Papendick. 1960. Worm worked soils of eastern South Dakota, their morphology and classification. Soil Science Society of America Proceedings 24:128-132.

Busscher, W. J., and R. E. Sojka. 1987. Enhancement of subsoiling effect on soil strength by conservation tillage. Transactions of the American Society of Agricultural Engineers 30:888-892.

Campbell, G. S. 1985. Soil physics with BASIC: Transport models for soil-plant systems. Amsterdam: Elsevier.

Campbell, R. B., D. C. Reicosky, and C. W. Doty. 1974. Physical properties and tillage of Paleudults in the southeastern Coastal Plains. Journal of Soil and Water Conservation 29:220-224.

Cannell, R. Q. 1977. Soil aeration and compaction in relation to root growth and soil management. Advances in Applied Biology 2:1-86.

Chaudhary, M. R., and S. S. Prihar. 1974. Root development and growth response of corn following mulching, cultivation, or interrow competition. Agronomy Journal 66:350-355.

Culley, J. L. B., W. E. Larson, and G. W. Randall. 1987. Physical properties of a Typic Hapludoll under conventional and no-tillage. Soil Science Society of America Journal 51:1587-1593.

Currie, J. A. 1966. The volume and porosity of soil crumbs. Journal of Soil Science 30:441-452.

Currie, J. A. 1984. Gaseous diffusion through soil crumbs: The effects of compaction and wetting. Journal of Soil Science 35:1-10.

Dexter, A. R. 1986. Model experiments on the behavior of roots at the interface between a tilled seed-bed and a compacted sub-soil. Plant and Soil 95:123-161.

Douglas, J. T., and M. J. Goss. 1987. Modification of porespace by tillage in two stagnogley soils with contrasting management histories. Soil and Tillage Research 10:303-317.

Douglas, J. T., M. G. Jarris, K. R. Howse, and M. J. Goss. 1986. Structure of a silty soil in relation to management. Journal of Soil Science 37:137-151.

Drew, M. C., and L. R. Saker. 1980. Direct drilling and plowing: Their effect on the distribution of extractable phosphorus and potassium, and of roots, in the upper horizons of two clay soils under winter wheat and spring barley. Journal of Agricultural Science (Cambridge) 94:411-423.

Edwards, C. A., and J. R. Lofty. 1977. Biology of earthworms. 2nd ed. New York: Halsted Press.

Edwards, W. M., J. B. Fehrenbacher, and J. P. Vavra. 1964. The effect of discrete ped density on corn root penetration in a Planosol. Soil Science Society of America Proceedings 28:560-564.

Edwards, W. M., L. D. Norton, and C. E. Redmond. 1988. Characterizing macropores that affect infiltration into non-tilled soil. Soil Science Society of America Journal 52:483-487.

Ehlers, W., U. Kopke, F. Hasse, and W. Bohm. 1983. Penetration resistance and root growth of oats in tilled and untilled loess soil. Soil and Tillage Research 3:261-275.

Elkins, C. B., and J. G. Hendrick. 1983. A slit-plant tillage system. Transactions of American Society of Agricultural Engineers 26:710-712.

Ellis, F. B., J. G. Elliott, B. T. Barnes, and K. R. Howse. 1977. Comparison of direct drilling, reduced cultivation and ploughing on the growth of cereals. Journal of Agricultural Science (Cambridge) 89:631-642.

Erickson, A. E. 1982. Tillage effects on soil aeration. In Predicting tillage effects on soil physical properties and processes. eds. P. W. Unger and D. M. Van Doren, Jr. 91-104. Special Publication Number 44. Madison, WI: American Society of Agronomy.

Erickson, A. E., and D. M. Van Doren. 1961. The relation of plant growth and yield to soil oxygen availability. In Transactions of the seventh international congress of soil science. ed. F. A. Baren, vol. 4:428-432. Amsterdam: Elsevier.

Fausey, N. R., and A. S. Dylla. 1984. Effects of wheel traffic along one side of corn and soybean rows. Soil and Tillage Research 4:147-154.

Gardner, B. L., and B. L. Blad. 1986. Evaluation of spectral reflectance models to estimate corn leaf area while minimizing the influence of soil background effects. Remote Sensing of the Environment 20:183-193.

Gerwitz, A., and E. R. Page. 1974. An empirical mathematical model to describe plant root systems. Journal of Applied Ecology 11:773-781.

Glinski, J., and W. Stepniewski. 1985. Soil aeration and its role for plants. Boca Raton, FL: Chemical Rubber Company Press.

Grable, A. R., and E. G. Siemer. 1968. Effects of bulk density, aggregate size, and soil water suction on oxygen diffusion, redox potentials, and elongation of corn roots. Soil Science Society of America Proceedings 32:180-186.

Greacen, E. L., K. P. Barley, and D. A. Farrell. 1968. The mechanics of root growth in soils with particular reference to the implications for root distribution. In Root growth. ed. W. J. Whittington, 256-269. London: Butterworth Publishers.

Green, R. E., L. R. Ahuja, and S. K. Chong. 1986. Hydraulic conductivity, diffusivity, and sorptivity of unsaturated soils: Field methods. In Methods of soil analysis: physical and mineralogical methods. 2nd. Ed., pt. 1. ed. A. Klute, 771-798. Madison, WI: American Society of Agronomy.

Grimes, D. W., W. R. Sheesley, and P. L. Wiley. 1978. Alfalfa root development and shoot regrowth in compact soil of wheel traffic patterns. Agronomy Journal 70:955-958.

Gupta, S. C., and R. R. Allmaras. 1987. Models to assess the susceptibility of soils to excessive compaction. Advances in Soil Science 6:65-100.

Gupta, S. C., and W. E. Larson. 1982. Modeling soil mechanical behavior during tillage. In Predicting tillage effects on soil physical properties and processes. ed. P. W. Unger and D. M. Van Doren, Jr., 151-178. Special Publication no.44. Madison, WI: America Society of Agronomy.

Gupta, S. C., P. P. Sharma, and S. A. DeFranchi. 1989. Compaction effects on soil structure. Advances in Agronomy 42:311-338.

Hackett, C., and D. A. Rose. 1972. A model of the extension and branching of a seminal root of barley and its use in studying relations between root dimensions. Australian Journal of Biological Science 25:1057-1064.

Hamblin, A. P. 1985. The influence of soil structure on water movement, crop root growth, and water uptake. Advances in Agronomy 38:95-158.

Hillel, D. 1980. Fundamentals of soil physics. New York: Academic Press.

Hillel, D., and H. Talpaz. 1976. Simulation of root growth and its effects on the pattern of soil water uptake by a nonuniform root system. Plant and Soil 52:325-343.

Horton, R., R. R. Allmaras, and R. M. Cruse. 1989. Tillage and compactive effects on soil hydraulic properties and water flow. In mechanics and related processes in structured agricultural soils, 187-204. Dordrecht: Kluwer Academic Press.

Jakobsen, B. F., and A. R. Dexter. 1987. Effect of soil structure on wheat root growth, water uptake and grain yield. A computer simulation model. Soil and Tillage Research 10:331-345.

Jakobsen, B. F., and A. R. Dexter. 1988. Influence of biopores on root growth, water uptake and grain yield of wheat (Triticum aestivum) based on predictions from a computer model. Biology and Fertility of Soils 6:315-321.

Johnson, M. D., B. Lowery, and T. C. Daniel. 1984. Soil moisture regimes in three conservation tillage systems. Transactions of the American Society of Agricultural Engineers 27:1385-1390,1395.

Johnson, W. C. 1962. Controlled soil cracking as a possible means of moisture conservation on wheat lands of the southwestern Great Plains. _Agronomy Journal_ 54:323-325.

Kirkham, D., and W. L. Powers, 1972. _Advanced soil physics_. New York: Wiley Interscience. Reprinted 1985. Malabar, FL: Krieger.

Klute, A. 1986. Water retention: laboratory methods. In _Methods of soil analysis: physical and mineralogical methods_. 2nd ed., pt. 1. ed. A. Klute, 635-662. Madison, WI: American Society of Agronomy.

Klute, A., and C. Dirksen. 1986. Hydraulic conductivity and diffusivity: Laboratory methods. In _Methods of soil analysis: physical and mineralogical methods_. 2nd ed., pt. 1. ed. A. Klute, 687-734. Madison, WI: American Society of Agronomy.

Kuchenbuch, R. O., and S. A. Barber. 1988. Significance of temperature and precipitation for maize root distribution in the field. _Plant and Soil_ 106:9-14.

Logsdon, S. D., J. C. Parker, and R. B. Reneau, Jr. 1987. Root growth as influenced by aggregate size. _Plant and Soil_ 99:267-275.

Logsdon, S. D., R. B. Reneau, Jr., and J. C. Parker. 1987. Corn seedling root growth as influenced by soil physical properties. _Agronomy Journal_ 79:221-224.

Lungley, D. R. 1973. The growth of root systems--a numerical computer simulation. _Plant and Soil_ 38:145-159.

McCoy, E. L. 1987. Energy requirements for root penetration of compacted soil. In _Future developments in soil science research_. ed. L. Boersma, 367-378. Madison, WI: Soil Science Society of America.

McInnes, K. J., E. T. Kanemasu, D. E. Kissel, and J. B. Sisson. 1986. Predicting diurnal variations in water content along with temperature at the soil surface. _Agricultural and Forest Meteorology_ 38:337-348.

McKeague, J. A., C. A. Fox, J. A. Stone, and R. Protz. 1987. Effects of cropping system on structure of Brookston clay loam in long-term experimental plots at Woodslee, Ontario. _Canadian Journal of Soil Science_ 67:571-584.

McKeague, J. A., C. Wang, and G. C. Topp. 1982. Estimating saturated hydraulic conductivity from soil morphology. _Soil Science Society of America Journal_ 46:1239-1244.

Misra, R. K., A. M. Alston, and A. R. Dexter. 1988. Root growth and phosphorus uptake in relation to size and strength of soil aggregates. Soil and Tillage Research 11:103-116.

Misra, R. K., A. R. Dexter, and A. M. Alston. 1986. Penetration of soil aggregates of finite size. Plant and Soil 94:59-85.

Moore, I. D. 1981. Effect of surface sealing on infiltration. Transactions of the American Society of Agricultural Engineers 24:1546-1561.

Narda, N. K., and R. B. Curry. 1981. SOYROOT - A dynamic model of soybean root growth and water uptake. Transactions of American Society of Agricultural Engineers 24:651-656.

Norman, J. M. 1979. Modeling the complete crop canopy. In Modification of the aerial environment of crops. ed. B. J. Barfield and J. F. Gerber, 249-277. Monograph 2. St. Joseph, MI: American Society of Agricultural Engineers.

Perrier, E. R., R. J. Millington, D. B. Peters, and R. J. Luxmoore. 1970. Wind structure above and within a soybean canopy. Agronomy Journal 62:615-618.

Philip, J. R., and D. A. de Vries. 1957. Moisture movement in porous materials under temperature gradients. Transactions of American Geophysical Union 38:222-228.

Pikul, J. L., Jr., and R. R. Allmaras. 1986. Physical and chemical properties of a Haploxeroll after fifty years of residue management. Soil Science Society of America Journal 50:214-219.

Prebble, R. E. 1970. Root penetration of smeared soil surfaces. Experimental Agriculture 6:303-308.

Rawls, W. J., D. L. Brakensiek, and B. Soni. 1983. Agricultural management effects on soil water processes. Transactions of American Society of Agricultural Engineers 26:1747-1752.

Rendig, V. V., and H. M. Taylor. 1989. Principles of soil-plant interactions. New York: McGraw Hill.

Richards, B. G., and E. L. Greacen. 1986. Mechanical stresses on an expanding cylindrical root analogue in granular media. Australian Journal Soil Research 24:393-404.

Russell, R. S. 1977. Plant root systems: their functions and interactions with soil. London: McGraw-Hill.

Russell, R. S., and M. J. Goss. 1974. Physical aspects of soil fertility, the response of roots to mechanical impedance. Netherlands Journal of Agricultural Science 22:305-318.

Schumacher, T. E., and A. J. M. Smucker. 1981. Mechanical impedance effects on oxygen uptake and porosity of drybean roots. Agronomy Journal 73:51-55.

Schumacher, T. E., and A. J. M. Smucker. 1984. Effects of localized anoxia on Phaseolus vulgaris L. root growth. Journal of Experimental Botany 35:1039-1047.

Shaffer, M. J., and W. E. Larson, eds. 1987. NTRM, soil crop simulation model for nitrogen tillage and crop residue management. Conservation Research Report 34-1. Agricultural Research Service, USDA, Washington, DC. 103 pp.

Sheppard, S. C., and M. H. Miller. 1977. Temperature changes and the geotropic reaction of the radicle of Zea mays L. Plant and Soil 47:631-644.

Shipitalo, M. J., and R. Protz. 1987. Comparison of morphology and porosity of a soil under conventional and zero tillage. Canadian Journal of Soil Science 67:445-456.

Smucker, A. J. M., and A. E. Erickson. 1989. Tillage and compactive modifications of gaseous flow and soil aeration. In Soil mechanics and related processes in structured agricultural soils, 205-222. Dordrecht: Kluwer Academic Press.

Soil Science Society of America. 1987. Glossary of soil science terms. Madison, WI. Soil Science Society of America.

Soil Survey Staff. 1951. Soil survey manual. Agricultural Handbook no. 18, USDA, Soil Conservation Service, Washington, DC. 503 pp.

Soil Survey Staff. 1975. Soil taxonomy: A basic system of soil classification for making and interpreting soil surveys. Agricultural Handbook no. 436, USDA, Soil Conservation Service, Washington, DC. 754 pp.

Southard, R. J., and S. W. Buol. 1988a. Subsoil blocky structure formation in some North Carolina Paleudults and Paleaquults. Soil Science Society of America Journal 52:1069-1076.

Southard, R. J., and S. W. Buol. 1988b. Subsoil saturated hydraulic conductivity in relation to soil properties in the North Carolina Coastal Plain. Soil Science Society of America Journal 52:1091-1094.

Tardieu, F. 1988. Analysis of the spatial variability of maize root density. Plant and Soil 107:259-266.

Tardieu, F., and H. Manichon. 1986. Characterization en taut que capteu d'eau de l'enracinement due mais en parcelle cultivée. Agronomie 6:18-37.

Tardieu, F., and H. Manichon. 1987. Etat structural, enracinement et alimentation hydrique du mais. *Agronomie* 7:201-211.

Taylor, H. M. 1983. Managing root systems for efficient water use: An overview. In *Limitations to efficient water use in crop production*. ed. H. M. Taylor, W. R. Jordon, and T. R. Sinclair, 87-113. Madison, WI: American Society of Agronomy,

Taylor, H. M., and B. Klepper. 1978. The role of rooting characteristics in the supply of water to plants. *Advances in Agronomy* 30:99-128.

van Genuchten, M. Th. 1978. *Numerical solutions of the one-dimensional saturated-unsaturated flow equation*. Research Report 78-WR-09. Water Resources Program, Department of Civil Engineering, Princeton University.

Van Stiphout, T. P. J., H. A. J. Van Lanen, O. H. Boersma, J. Bouma. 1987. The effect of bypass flow and internal catchment of rain on the water regime in a clay loam grassland soil. *Journal of Hydrology* 95:1-11.

Voorhees, W. B. 1989. Root activity related to shallow and deep compaction. In *Mechanics and related processes in structured agricultural soils*, 173-186. Dordrecht: Kluwer Academic Press.

Voorhees, W. B., R. R. Allmaras, and C. E. Johnson. 1981. Alleviating temperature stress. In *Modifying the root environment to reduce crop stress*. ed. G. F. Arkin and H. Taylor, 217-266. Monograph 4. St. Joseph, MI. American Society of Agricultural Engineers.

Voorhees, W. B., D. A. Farrell, and W. E. Larson. 1975. Soil strength and aeration effects on root elongation. *Soil Science Society of America Proceedings* 39:948-953.

Voorhees, W. B., W. W. Nelson, and G. W. Randall. 1986. Extent and persistence of subsoil compaction caused by heavy axle loads. *Soil Science Society of America Journal* 50:428-433.

Walker, J. M. 1969. One-degree increments in soil temperature affect maize seedling behavior. *Soil Science Society of America Proceedings* 33:729-736.

Wang, J., J. D. Hesketh, and J. T. Woolley. 1986. Preexisting channels and soybean rooting patterns. *Soil Science* 141:432-437.

Whisler, F. D., J. R. Lambert, and J. A. Landivar. 1982. Predicting tillage effects on cotton growth and yields. In *Predicting tillage effects on soil physical properties and processes*. ed. P. W. Unger and D. M.

Van Doren, 179-198. Special Publication no. 44. Madison, WI: American Society of Agronomy.

Whiteley, G. M., and A. R. Dexter. 1983. Behavior of roots in cracks between soil peds. _Plant and Soil_ 74:153-162.

Whiteley, G. M., and A. R. Dexter. 1984. Displacement of soil aggregates by elongating roots and shoots of crop plants. _Plant and Soil_ 77:131-140.

Wierenga, P. J., and C. T. deWit. 1970. Simulation of heat flow in soils. _Soil Science Society of America Proceedings_ 34:845-848.

C. E. Clapp
J. A. E. Molina
R. H. Dowdy

3   Soil Organic Matter, Tillage, and the Rhizosphere

Agricultural production systems that control and manage soil organic matter levels can directly influence the environment and ecology of the rhizosphere. Management of crop residues by return to the soil surface, incorporation, or removal, coupled with appropriate tillage practices, can affect all the environmental factors controlling biological activities of a surface soil. Abundant information is available on organic matter sources, contents, composition, properties, and interactions. Of particular significance are the carbon (C) and nitrogen (N) cycles in the soil-water-plant system. In this chapter, we will consider the interrelationships among soil organic matter, carbon and nitrogen transformations as influenced by temperature and water, and tillage and residue management in the rhizosphere of mineral soils. The use of simulation models can provide an excellent means of integrating these concepts and information into efficient systems designed to manage agricultural and environmental problems.

SOIL ORGANIC MATTER

General Considerations

Organic matter in mineral surface soils exhibits both direct and indirect effects on the rhizosphere. To evaluate these effects, we must consider amounts and distribu-

tion, composition, functions, and environmental significance of soil organic matter. The phrase "soil organic
matter" is used to refer specifically to the nonliving soil
components, which are a heterogeneous mixture composed
largely of products resulting from microbial and chemical
transformations of organic residues. Many general (Kononova
1966; Allison 1973; Gieseking 1975; Schnitzer and Khan
1978) and specific (Hayes and Swift 1978; Stevenson 1982,
1986; Hayes et al. 1989) reviews on soil organic matter and
humic substances are available. Here we will attempt only
to summarize the more important contributions and interactions of soil organic matter and its components by relating their properties to rhizosphere dynamics and ecology.

Soils vary widely in organic matter content. Variations in amounts of organic matter in mineral soils make it
difficult to provide representative values. The organic
matter content in the surface 15 to 20 cm depth may range
from a trace to 15 to 20 percent. The ranges and average
organic matter contents (and related nitrogen values) for a
representative number of U.S. soils are shown in Table
3.1. The widest range is for West Virginia (0.7 to 15
percent) and the narrowest is for Minnesota (3.4 to 7.4
percent). Typical surface prairie grassland soils
(Mollisols) may contain 5 to 6 percent organic matter; some
sandy soils (Haploborolls), less than 1 percent. Poorly
drained soils (Aquepts) often have organic matter contents
higher than 10 percent. In contrast, tropical soils
(Oxisols) usually have low amounts of organic matter. As
might be expected, subsoils, below 15 to 30 cm in depth,
generally have much lower organic matter contents than
surface soils (Table 3.2). Organic matter accumulates in
the upper layers because most of the organic residues in
both cultivated and virgin soils are incorporated in, or
are returned to the surface.

Amounts and distribution of soil organic matter are
regulated by the so-called "factors of soil formation,"
that is, time, climate, vegetation, parent material, and
topography (Stevenson 1982). Organic matter does not accumulate indefinitely in well-drained soils but attains equilibrium with time that is governed by the other four factors. Climate probably is the single most important factor, since it controls plant ecology, the quantity of plant
material produced, the decomposition rate, and the intensity of microbial activity in the soil. The influence of
vegetation is highlighted by the difference between grassland soils and forest soils, with much higher organic matter synthesis in the rhizosphere under grass. Parent mate-

rial is effective mainly through textural properties, with fine-textured soils having higher organic matter contents than loamy soils, which in turn have higher organic matter contents than sandy soils. Topography is a factor in organic matter content through influence on climate and water-related conditions, where micro-relief and aerobic/anaerobic situations can determine decomposition rates and activity of microorganisms. Another factor affecting organic matter contents and distribution and often superimposed on the soil-formation factors is soil management (Follett, Stewart, and Cole 1987), including tillage, fertilization, residue return, and cropping sequences.

The general composition of soil organic matter, including extraction, fractionation, and characterization techniques, has been studied extensively for over 200 years (Stevenson 1982). Early workers attempted to isolate and

TABLE 3.1 Organic matter and nitrogen contents of mineral surface soils from several U.S. locations.[†]

Soil location	Organic matter		Nitrogen	
	Range	Mean	Range	Mean
		------%------		
Kansas (117)[‡]	0.1-3.6	3.3	0.02-0.27	0.17
W. Virginia (240)	0.7-15	2.9	0.04-0.54	0.15
N. Carolina (52)	0.8-1.5	1.2	0.04-0.08	0.06
S. Great Plains (21)	1.2-2.2	1.6	0.07-0.14	0.10
Utah (21)	1.5-4.9	2.7	0.09-0.26	0.15
Pennsylvania (15)	1.7-9.9	3.6	0.08-0.50	0.18
Nebraska (30)	2.4-5.3	3.8	0.12-0.25	0.18
Minnesota (9)	3.4-7.4	5.2	0.17-0.35	0.27

[†] Adapted from Brady 1984.
[‡] Numbers in parentheses indicate number of soils.

purify individual chemical compounds of specific composition; by the end of the nineteenth century, however, it
had been established that humus (now synonymous with soil
organic matter) was a complex mixture of weakly acidic
organic substances that were mostly colloidal in nature.
During the early twentieth century, renewed efforts were
made to classify humic substances and determine their
chemical nature and structure. Chemists in the coal and
beer industries related their products to soil organic substances, both in origin and composition. Waksman (1938)
regarded humus as a mixture of plant-derived materials,
including fats, waxes, resins, hemicellulose, cellulose,
and "ligno protein," and his "proximate analysis" method
was widely used on mineral soils, peats, composts, and
organic layers of forest soils. Today, humus is known to
include a broad spectrum of organic constituents, many of
which have counterparts in plant and animal tissues. The
two major types of compounds identified are nonhumic substances, consisting of well-known organic constituents such
as amino acids, carbohydrates, and lipids; and humic substances, a series of high-molecular-weight, brown-to-black
materials formed by secondary synthesis reactions, which
are distinctive to soils, sediments, and aquatic environments. A more detailed discussion of humic substances will
follow in another section of this chapter.

The significance of organic matter in soils can be
placed into proper perspective when we consider its contri-

TABLE 3.2 Distribution of organic matter in four soil
profiles.[†]

| Soil depth | Minnesota prairie soil | | Indiana forest soil | |
	Well drained	Poorly drained	Well drained	Poorly drained
cm	------------% Organic matter----------			
0-15	7.0	8.5	4.0	13.0
15-30	6.0	7.0	1.5	2.5
30-45	5.0	5.0	1.0	1.5
45-60	4.0	4.0	0.7	1.5
60-75	3.0	3.0	0.7	1.5
75-90	2.0	2.0	0.7	1.5

[†] Adapted from Brady 1984.

bution to physical, chemical, and biological properties of the soil. Physically, humus promotes aggregation of primary soil particles, thereby improving tilth, aeration, and movement and retention of water, and influences soil temperature due to its dark color. Chemically, it serves as a source of nitrogen, phosphorus, sulfur, and trace metals for plant growth as a result of decomposition and mineralization. In addition, humus provides exchange sites for buffering and nutrient-holding capacity, and enhances solubility and availability of minerals through chelation mechanisms. Biologically, humus affects activities of microflora and microfauna by serving as a source of energy for bacteria, actinomycetes, fungi, and earthworms and possibly acts as a growth-promoting substance for higher plants.

The fourth important effect of soil organic matter on rhizosphere ecology is one of environmental significance. Humic materials are significant components of sewage and wastewater treatment systems, with land application of municipal sewage sludges and wastewater effluents providing an efficient alternative for utilization of these wastes (Page et al. 1983; Clapp et al. 1984; Clapp et al. 1986; Boyd and Sommers 1990). Soil organic matter interactions with pesticides and other toxic organic compounds can influence phytotoxicity, leachability, volatility, and biodegradability of these potential pollutants. Composting organic wastes is beneficial when these materials are added to soil. Humic substances are often found also in drinking water supplies and must be removed for both aesthetic and health reasons.

Soil Nitrogen

Generally, the organic matter content of most arable soils is directly related to the nitrogen content, and considering the constant C:N ratio of 10:1 to 12:1 for organic matter, the ratio between organic matter and nitrogen is about 20:1 (Brady 1984). Nitrogen values for several U.S. surface soils are given in Table 3.1. The importance of the organic matter to nitrogen relationships will be readily apparent in later discussions of carbon and nitrogen cycling and transformations.

The surface layer of most mineral soils contains more than 90 percent of the nitrogen in organic forms, with the rest occurring as exchangeable or fixed NH_4^+. Agricultural soils in the United States have values between 0.02

and 0.5 percent nitrogen (Table 3.1). Extensive reviews on the chemistry of soil organic N include those of Stevenson (1982), Schnitzer (1985), Anderson et al. (1989), and Stevenson and He (1990). Standard procedures for fractionating and quantitatively identifying organic nitrogen compounds are shown in Table 3.3. The more important forms of nitrogen in the rhizosphere for microbial and plant nutrients are the 55 to 80 percent of the total, occurring

TABLE 3.3 Fractionation of soil nitrogen following acid hydrolysis.[†]

Form of nitrogen	Definition and method	Soil N range
		---%---
Acid insoluble-N	Nitrogen remaining in soil residue following acid hydrolysis. Usually obtained by difference (total soil N minus hydrolyzable N).	20-35
NH_3-N	Ammonia recovered from hydrolysate by steam distillation with MgO.	20-35
Amino acid-N	Usually determined by the ninhydrin-CO_2 or ninhydrin-NH_3 methods. Recent workers have favored the latter.	30-45
Amino sugar-N	Steam distillation with phosphate-borate buffer at pH 11.2 and correction for NH_3-N. Colorimetric methods can also be used. Sometimes referred to as hexosamine-N.	5-10
Hydrolyzable unknown-N (HUN fraction)	Hydrolyzable N not accounted for as NH_3, amino acids, or amino sugars. Part of this N occurs as non-α-amino-N of arginine, tryptophan, lysine, and proline.	10-20

[†] Adapted from Stevenson 1982.

as NH_3, amino acid, and amino sugar nitrogen. Data illustrating the effect of cultivation and cropping systems on distribution of forms of nitrogen in soil are shown in Table 3.4. Intense cultivation leads to a general decline in total nitrogen content but a slight increase in the proportion of nitrogen as hydrolyzable NH_3. In contrast, the proportion of soil nitrogen as amino acid nitrogen generally decreases with cultivation, while amino sugar nitrogen changes very little. The conclusion that cultivation has little influence on soil nitrogen emphasizes the fact that all nitrogen forms are biodegradable and that separation of soil nitrogen hydrolyzates into components would not be practical as an index of nitrogen availability nor as a predictor of crop yield.

The high stability of soil organic nitrogen complexes in resisting microbial attack is of great importance to the nitrogen balance of the soil. Several reasons often advanced to explain this stability are: (1) resistant materials are formed by reaction of proteinaceous compounds (e.g., amino acids, peptides, proteins) with other organic components of soil such as reducing sugars, quinones, tannins, and lignins; (2) chemical combinations involving NH_3 or NO_2 with humic substances or lignin produce biologically re-

TABLE 3.4   Effect of cultivation and cropping system on distribution of the forms of nitrogen in soil.[†]

Location & treatment	Acid insoluble	NH_3	Amino acid	Amino sugar	HUN
			%		
Illinois (2)[‡]					
grass	20.3	16.6	42.0	10.5	10.7
cultivated	20.2	16.7	35.0	14.4	13.9
Iowa (10)					
virgin	25.4	22.2	26.5	4.9	21.0
cultivated	24.0	24.7	23.4	5.4	22.5
Nebraska (4)					
virgin	20.8	19.8	44.3	7.3	7.8
cultivated	19.3	24.5	35.8	7.0	13.4

Column group header: Form of N

[†] Adapted from Stevenson 1982.
[‡] Numbers in parentheses indicate number of soils.

sistant complexes; (3) adsorption of organic nitrogen compounds by soil clay minerals protects the complex from degradation; (4) reactions between organic nitrogen constituents and polyvalent cations such as Fe^{3+} and Al^{3+} form complexes that are resistant to biological attack; and (5) physical inaccessibility of some of the organic nitrogen compounds within small soil pores shields these constituents from bacterial or even enzyme attack.

The dynamics of soil nitrogen involve a wide variety of transformations, most of which include the organic fraction. The biological interactions of nitrogen through mineralization/immobilization provide the main transfer of organic residues into stable humus materials. Biochemical processes such as ammonification, nitrification, denitrification, and assimilation often go on simultaneously and are responsible for the turnover of nitrogen within the soil. Chemical reactions involving inorganic nitrogen and soil organic matter may also be of significance. Techniques using the labeled nitrogen isotope (^{15}N) as a tracer now can provide new contributions in studying soil nitrogen transformations; that is, turnover and stabilization of nitrogen, including incorporation of fertilizer nitrogen into various soil organic matter fractions and nitrogen pools (see reviews on ^{15}N research by Hauck and Bremner 1976; Bremner and Hauck 1982).

Stevenson (1982) predicted that future agronomic management practices would be geared toward conserving energy and minimizing adverse environmental effects of nitrogen fertilizers. This picture will be facilitated by computer simulation models of nitrogen dynamics in soils.

Soil Humic Substances

Humic substances are among the most widely distributed organic materials on earth. According to Stevenson (1982), the amount of carbon occurring as humic acids (60 x 10^{11} Mg) is greater than that in living organisms (7 x 10^{11} Mg). Because of their abundance in surface soils, humic substances and their interactions with other soil components are of prime importance to the environment of the rhizosphere. By recent definition of the International Humic Substances Society (IHSS), soil humic substances are amorphous, polymeric, yellow- to brown-colored components of soil organic matter, composed of fulvic acids, soluble in aqueous alkali and acid; humic acids, precipitated when alkali-soluble components are acidified to pH 1; and humins, insoluble in aqueous alkali and acid. Several excellent reviews of the biochemistry and functions of soil humic substances have been prepared by Hayes and Swift (1978), Stevenson (1982), Hayes (1984), Aiken et al. (1985), Orlov (1985), and MacCarthy et al. (1990).

A modern view of the formation of humic substances, according to Stevenson (1982), suggests that humic and fulvic acids are formed by a multistage process that includes decomposition of all plant components, including lignin, into simpler monomers; metabolism of monomers with an accompanying increase in the soil biomass; repeated recycling of the biomass carbon and nitrogen with synthesis of new cells; and concurrent polymerization of the reactive monomers into high-molecular-weight polymers. The general consensus is that polyphenols (quinones) synthesized by microorganisms, together with those liberated from lignin, polymerize alone or in the presence of amino compounds (amino acids) to form brown-colored polymers. A suggested Maillard or "browning" reaction consisting of the condensation of amino acids with reducing sugars may be active under some northern temperate climatic conditions.

To carry out meaningful characterizations and reaction studies of humic substances, it is first necessary to extract the components and to fractionate them into reasonable, representative entities. The more recent approach has been to use aqueous neutral or alkaline salt solutions such as $Na_4P_2O_7$ alone or combined with NaOH to isolate the maximum amounts of humic material, and to fractionate this material employing standard procedures of the IHSS (Aiken et al. 1985). The most comprehensive treatment of extraction and fractionation procedures for soil humic substances has been presented by Hayes (1985) and Swift (1985), respectively.

Chemical characterization of soil humic substances has included elemental and functional group analysis as well as determination of chemical structures using various degradative and nondegradative techniques (Stevenson 1982). Table 3.5 shows a range of carbon, oxygen, hydrogen, and nitrogen contents for various extracting solvents. The carbon content of humic acids ranges from 51 to 56 percent; oxygen content varies from 33 to 41 percent. Fulvic acids have lower carbon contents (37 to 55 percent), but higher oxygen (37 to 51 percent). Reactive functional group analyses have focused on oxygen-containing groups, including COOH, phenolic- and enolic-OH, alcoholic-OH, and C=O of quinones, hydroxy-quinones, and α,β-unsaturated ketones. Total acid-

TABLE 3.5   Carbon, oxygen, hydrogen, and nitrogen contents of humic acids (HA) and fulvic acids (FA) isolated by various solvents.[†]

Solvent	C		O		H		N	
	HA	FA	HA	FA	HA	FA	HA	FA
				%				
Pyridine	55.9	47.1	32.9	39.9	5.1	5.3	4.4	6.0
Dimethyl-formamide	54.3	52.3	36.8	38.8	4.6	4.1	2.6	3.2
Dimethyl-sulfoxide	55.0	55.0	35.5	37.1	4.2	4.4	3.3	2.2
Na-EDTA	52.1	48.4	--	--	4.1	4.2	--	--
0.5M NaOH	53.1	45.0	36.3	43.0	6.0	6.0	2.9	4.3
0.1M $Na_4P_2O_7$ (pH7)	50.9	37.3	41.1	50.9	3.3	5.1	3.0	5.0

[†] Adapted from Hayes 1985. Values on an ash-free basis.

ity (approximately equal to cation exchange capacity) of fulvic acids (6.4 to 14.2 meq g^{-1}) is generally higher than for humic acids (5.6 to 7.7 meq g^{-1}). Degradation methods show great promise for characterizing humic substances (Stevenson 1982), provided a combination of degradative procedures with increasing severity is used. For the primary structure of humic acids, contemporary investigations favor a "type" molecule consisting of polymeric micelles, the basic structure of which is an aromatic ring of the di- or trihydroxy phenol type bridged by -O-, -CH$_2$-, -NH-, -N=, and -S- groups and containing both free OH groups and the double linkages of quinones. In nature, the molecule may contain attached proteinaceous and carbohydrate residues.

Newer methods of isolation and characterization involving supercritical gas extraction techniques coupled with pyrolysis/mass spectrometry show much promise (Schnitzer 1990). Nondegradative procedures have added significantly to knowledge of structures and functional groups in humic substances (Hayes 1984). Especially promising are spectroscopic techniques, such as visible, ultraviolet (UV), and infrared (IR) methods (particularly the Fourier transform type, FTIR); electron spin resonance (ESR); electron nuclear double resonance (ENDOR); and nuclear magnetic resonance (NMR, both ^{13}C-NMR and ^{15}N-NMR types), with special modifications of cross polarization (CP) and magic angle spinning (MAS) giving high resolution for solid-state samples. Spectroscopy will prove useful for indicating the type of linkages (secondary structure) between the primary humic repeating units.

The tertiary structures of soil humic substances, defined as the size, shape, and arrangement in space of the components, is important because their colloidal properties can affect the reactivity of the macromolecules. Particle size and shape can be determined by a combination of biochemical and physico-chemical methods, including ultracentrifugation, viscosity, vapor pressure lowering, light scattering, electron microscopy, x-ray diffraction and scattering, and gel and membrane filtration (Hayes et al. 1989). The average range in molecular weights measured for humic acids is 1,300 to 26,000 daltons (number average) and 30,000 to 300,000 daltons (weight average); for fulvic acids, the range is only 500 to 2,000 daltons (Orlov 1985). The modern view of the colloidal nature of humic acids is that they are coiled, long-chain macromolecules or two- or three-dimensional cross-linked macromolecules whose size and shape in solution are influenced by pH and neutral salts.

Several specific properties of humic substances can be directly related to the environment in the rhizosphere of agricultural surface soils (Frimmel and Christman 1988). These interactive properties include trace metal binding with humic acids and solubilization and movement with fulvic acids; adsorption of humic macromolecules on clay minerals to form crumbs or aggregates; reaction of pesticides with humic substances, the most active fraction of soil components, to promote nonbiological degradation, reduce phytotoxicity to subsequent crops, and prevent contamination of surface and groundwaters; and biotic effects between humic substances and plants (Chen and Aviad 1990) and microorganisms, by interaction with enzymes as inhibitory or stimulatory agents.

Root Contributions

The organic material released by roots into the surrounding soil volume, loosely and collectively referred to as "root exudates," is an extremely important component of the rhizosphere ecosystem. Physical and environmental factors causing the release of carbon from roots include temperature and pH extremes, light intensity, plant injury, and soil water content. The following compounds have been reported by Rovira and McDougall (1967) to be contained in wheat root exudates: sugars, amino acids, organic acids, nucleotides, and enzymes. These constituents probably act mainly as a food source to support large populations of microorganisms, but they also have important effects on soil properties and nutrient availability. In a nonsterile root system, as much as 20 percent of the dry matter assimilated by wheat plants was estimated by Martin (1977) to be lost as root exudates. The amount added to the rhizosphere can be as much as 50 to 100 mg of organic matter per gram of root per day . Only minor attention has been paid to the contributions of root exudates and other root-related debris and residues, with respect to the biochemical reactions of soil organic matter and humic substances. Sloughed root material, often referred to as lysate rather than exudate, and the microbial polysaccharides intimately associated with the root surfaces may function in soil aggregate stabilization and ion exchange. Some appropriate references to the plant root-soil interface and soil organic matter interactions include articles and reviews by Rovira and McDougall (1967), Rovira and Davey (1974), Harley and Russell (1979), Foster, Rovira, and Cock (1983), Todd and Giddens (1984), and Curl and Truelove (1986).

CARBON AND NITROGEN CYCLES IN THE RHIZOSPHERE

The dynamics of carbon and nitrogen in the rhizosphere are not completely known. Common wisdom has it that the activity must be more intense than in the nonrhizosphere portion of the soil, which is not the immediate recipient of the carbon (energy) and nitrogen from root exudates and sloughed-off materials.

The transmission electron micrographs of the root-soil interface (Foster, Rovira, and Cock 1983) provide compelling evidence of the dominant role that microbes must play in the rhizosphere, both regarding C-N biomass and as agents of biochemical transformations. The direct determination of carbon and nitrogen compounds and their behavior in the rhizosphere calls for special sampling techniques: physical separations of soil firmly adhering to roots, fal-

low versus cropped soils, or nutrient transfer in mono-
versus intercropping systems.

Much of the information on quantities of root exudate
and sloughed-off material has been obtained from hydroponic
experiments. Furthermore, interpretations have been based
on defining the rhizosphere as soil which firmly adheres to
roots (Rovira 1969; Rovira, Foster, and Martin 1979).
Recent publications have given the rhizosphere another oper-
ational definition. Helal and Sauerbeck (1983) devised a
plant growth unit with three soil zones separated by vertical steel mesh plates. Similarly, Klemedtsson et al.
(1987) distinguished four soil vertical layers by inter-
posing nylon mesh nets in a growth chamber; seeds were
planted in the inner layer, which contains the highest root
density. These separation techniques coupled with the use
of tracer carbon and nitrogen have shown that rhizodeposits
can be found beyond the diffusion range of soluble exudates, 30 mm away from roots, lending to the speculation
that volatile organics may be released from roots (Helal
and Sauerbeck 1987). These techniques, however, have not
been tried in situ.

Carbon Transformations

The partition of photoassimilates between the components of the soil/plant system has been reviewed by Lambers
(1987). The carbon balance sheet for the soil-root system
must account for residual root and rhizodeposits as well as
whatever has been microbially mineralized from these two
categories and lost as CO_2. This measurement calls for a
plant growth unit that enables one to separate $^{14}CO_2$ supplied to the shoot from $^{14}CO_2$ respired by the soil and the
roots. Meharg and Killham (1988), who compared carbon root
input to soil from ryegrass tagged with either a single
$^{14}CO_2$ pulse or prelabeled with a series of $^{14}CO_2$ pulses,
found exudates and respired CO_2 coming from recently
fixed carbon. Paul and Kucey (1981) applied a 48-hour
pulse of $^{14}CO_2$ to five-week-old Faba beans, which were
then given 96 hours to distribute the ^{14}C photosynthates in
roots, symbionts (mycorrhizal fungus or nitrogen-fixing
bacteria), and soil. The bulk of the assimilates were
respired by the roots (22 to 33 percent), 4 to 19 percent
flowed through the symbionts, and only 0.5 percent was
translocated to the rhizosphere. When no distinction is
made between mycorrhizal biomass and rhizodeposition,
measured fractions of assimilates translocated to the soil

are usually higher than 0.5 percent, and in the order of Paul's data for exudates plus mycorrhizal fungi, 7 to 60 percent of the photoassimilates translocated (Table 3.6) (Barber and Martin 1976; Johnen and Sauerbeck 1977; Whipps 1984; Haller and Stolp 1985; Milchunas et al. 1985; Helal and Sauerbeck 1987; Trofymow, Coleman, and Cambardella 1987; Biondoni, Klein, and Redente 1988; Davenport and Thomas 1988). Under axenic conditions, Trofymow, Coleman, and Cambardella (1987) found that 26-day-old oat plants had deposited in the rhizosphere 50 percent of the standing root biomass carbon in a sand-zeolite-$CaCO_3$ mixture.

The influence of roots on the turnover of the soil organic matter has been measured by the rate of degradation

TABLE 3.6   Carbon input from roots.[†]

Plant	Days after seeding	Soil	Root C	
			Ratio of rhizo-depositions to residuals	Ratio of total assimilates minus residuals to residuals[‡]
Corn	30	Loamy silt	0.11	0.51
Wheat	30	Sandy	0.08	0.48
	98	Sandy	0.22	0.37
	119	Sandy	0.18	2.1
	153	Sandy	0.20	3.2
Mustard	73	Sandy	0.57	2.7
	73	Sandy	0.60	2.1
	73	Loamy sand	0.34	1.4
	73	Sandy loam	0.60	1.9

[†] Adapted from Sauerbeck and Johnen (1977) and Helal and Sauerbeck (1987).

[‡] Total assimilates include residuals, or roots mechanically removed from the soil, rhizodepositions (organic metabolites and microbial biomass from root carbon input), and 70 percent of the carbon dioxide evolved from the root-soil system.

of ^{14}C-labeled organic matter in fallow versus cropped soils. Results have shown stimulatory (Billes and Botner 1981) as well as inhibitory effects (Jenkinson 1977; Sparling, Cheshire, and Mundie 1982). Reid and Goss (1983) observed more ^{14}CO$_2$ evolution but less ^{14}C in cold-water extracts from fallow soils than in soils planted with maize and perennial ryegrass; some ^{14}C translocated into the plants. Overall, soil organic matter decomposition was not quantitatively influenced by the plants, although an effect could be found for some chemically defined organic matter fractions. The destabilizing role of roots on soil organic matter has been well documented by Helal and Sauerbeck (1987). They consider that root exudates favor the build up of decomposers that attack organic matter exposed by root disruption of soil aggregates. Stabilization of soil organic matter by root exudation has also been documented. Legg et al. (1971) have shown that incorporation of mineral ^{15}N into the organic matter was accelerated by a series of oat crops.

Nitrogen Transformations

The release of nitrogen in soil from roots is not as well documented as that of carbon. A few data are available: Poth, Favre, and Focht (1986) have estimated that roots of Cajanus cajan (pigeon pea), a nitrogen-fixing plant, released in soil from 144 to 179 kg N ha^{-1}. Measurements were made by the soil nitrogen dilution technique, whereby nitrogen input from roots is quantified by the extent to which it dilutes the ^{15}N-enriched content of prelabeled soil organic matter. When the technique was applied to alfalfa, no transfer of nitrogen from roots to soil could be detected (Lory, Russelle, and Heichel 1988). The release of nitrogen in the rhizosphere has also been documented by measuring the amount of nitrogen symbiotically fixed and transferred to nonfixing plants grown in association (Eaglesham et al. 1981; Kang, Wilson, and Spikens 1981; Brophy, Heichel, and Russelle 1987; Sanginga, Mulongoy, and Ayanaba 1988; Papastylianou 1988). Patra, Sachdev, and Subbiah (1986) have shown that in the intercropping wheat-gram (Cicer arietum) and maize-cowpea systems, the cereal received from 14 to 32 percent of its nitrogen from the associated legumes. Surprisingly, the intercropping system turned out to be truly symbiotic, as legumes fixed more nitrogen than when grown alone; this effect was fertilizer-nitrogen dependent. In contrast,

Ofori, Pate, and Stern (1987) could not measure any nitrogen transfer from cowpea to maize during intercrop growth.

The methodology to measure nitrogen transfer between plants grown in association calls for particular care. Ta, Faris, and Macdowell (1989) have compared four techniques: direct ^{15}N transfer from atmospheric N_2 to fixing plant, then to nonfixing plant; the leaf-labeling method, whereby the fixing plant is prelabeled by exposing it to nitrate ^{15}N; the plant nitrogen dilution method, which consists of prelabeling both plants with ^{15}N and comparing the extent of ^{15}N dilution in the nonfixing plant grown alone and in association with the fixing plant; and the traditional difference or total nitrogen balance method. With the exception of the total nitrogen balance approach, which overestimated nitrogen transfer, the other three methods gave comparable results: a transfer of 3 percent of fixed nitrogen in alfalfa to timothy (Phleum pratense), corresponding to 10 percent of total nitrogen in timothy. Boller and Nosberger (1988) have shown that nitrogen-transfer measurements could be affected by differences in the rooting pattern between ryegrass grown alone or in a mixture with clover. The beneficial association of intercropping is not limited to nitrogen-fixing plants. Fisher and Stone (1969) have obtained evidence indicating that higher nitrogen content in the herbaceous vegetation growing under pine or larch plantations was caused by increased nitrogen availability in the conifers' rhizosphere.

Nitrification and denitrification have been reported to be both inhibited and stimulated in the rhizosphere (Rice and Pancholy 1972; Woldendorp 1975; Smith and Tiedje 1979; Berg and Rosswall 1987). Klemedtsson et al. (1987) found more nitrite than ammonium oxidizers in the rhizosphere of barley; they proposed that conditions in the rhizosphere favor a cycling between nitrate reduction and nitrite oxidation. Symbiotic and non-symbiotic atmospheric nitrogen fixation is not considered in this chapter.

Carbon-Nitrogen Models

An appreciation of the carbon dynamics in the rhizosphere can be obtained with simple conceptual and mathematical models driven by a few key experimental data. Thus Wood (1987), starting from experimental estimates of barley root weight and rates of loss, computed root loss to root weight ratios of 1.0 and 2.5 for 77- and 189-day-old bar-

ley, respectively. This was of the order of the values
found by Helal and Sauerbeck (1987) for wheat (Table 3.6).
A more elaborate model of microbial and carbon flux in the
rhizosphere was developed by Newman and Watson (1977). The
flux of substrate in the rhizosphere was computed as the
sum of substrate flow by diffusion, through the microbial
biomass, and from the soil organic matter. The boundary
conditions were the rate of substrate exudation at the root
surface and substrate equilibrium at the interface between
the soil and the rhizosphere. The model was used to
simulate the effect of soil water content, root density,
root exudation rate, and microbial growth constants on the
microbial and substrate concentrations in the rhizosphere.

SOIL TILLAGE AND MANAGEMENT

The influence of soil tillage on the rhizosphere is
indirect. On the one hand, tillage operates on the scale
of centimeters, while the rhizosphere does not extend more
than one to two millimeters beyond the root surface. This
zone (the rhizosphere) is characterized and dominated by
intensified microbial activity that is dependent upon the
living root and associated degradation products (Curl and
Truelove 1986). Hence, tillage directly affects the physical, chemical, and biological factors that, taken as a
whole, represent the environment surrounding the rhizosphere.

Tillage Interactions

Tillage, through its impact on soil structure, significantly affects water and heat transfer in the soil and gaseous exchange with the atmosphere. Allmaras and Logsdon
(1990) have presented an excellent analysis that documents
the effects of soil tillage on soil structure. Because
soil organic matter serves as the chief binding agent for
aggregate stability, its influence on root development and
rhizosphere dynamics is large. Long-term cultivation of
croplands results in a reduction of total soil organic
matter (Doran and Smith 1987). One of the main consequenc-
es of this destruction of soil organic matter is the mineralization of carbon and nitrogen discussed earlier in this
chapter. Losses of carbon and nitrogen can be substantial,
as noted by Tiessen and Stewart (1983), who observed net

losses of 34 percent carbon and 29 percent nitrogen over sixty years of cultivating Cryoborolls under small grain-fallow rotation. Under tropical conditions, losses of carbon and nitrogen are even greater (Lal, DeVleeschauwer, and Nanje 1980). These losses can be reversed by moving from a moldboard-based tillage system into a conservation or no-tillage system (Blevins, Fryer, and Smith 1985; Gallaher and Ferrer 1987).

The destruction of soil organic matter and associated degradation of soil structure leads to soil compaction and increased mechanical resistance to root development (Hamblin 1985). When cultivated soils are returned to zero-tillage, bulk density of the 0 to 20 cm Ap horizon increases without substantial reduction in root growth (Ehlers 1982). It was suggested that roots utilize the biopores of nontilled soil, and, hence, no reduction in root growth occurs. Recent reviews by Hamblin (1985) and Allmaras and Logsdon (1990) do an excellent job of discussing the interrelationships between tillage-induced soil compaction and water and heat fluxes on root development. Placement of crop residues in and on the soil surface by tillage operations is one of the most readily available management tools for optimizing soil water and temperature regimes for root growth (Doran and Smith 1987).

Residue Effects

The role of crop residues and organic manure amendments on rhizosphere biology is an open question and may be more indirect than direct in nature. Curl and Truelove (1986) noted that most evidence suggests that organic manures have very little influence on rhizosphere flora, but instead directly affect plant growth, which in turn influences root exudation and associated microbial activity. Elliott, Papendick, and Bezdicek (1987), however, reported poor growth and overwinter stand losses in no-till seeded (surface residue) winter wheat and suggested that the problem was associated with root colonization by inhibitory pseudomonads during cold, wet springs. In contrast, Herman (1985) reported greater antagonistic activity of rhizosphere mycoflora against Gaeumannomyces graminis under zero-tillage than under conventional tillage. Allmaras, Kraft, and Miller (1988) have published a thorough review on the effects of soil compaction and crop residue management on root health, while Hunt (1990) has reviewed the broader issue of microbiological dynamics in the rhizosphere.

Abandonment of conventional tillage for a reduced or no-tillage system results in a concentration of roots near the soil surface (Dowdy et al. 1988). Associated with conversion to a no-tillage system is the acidification of the near-surface soil layer (Thomas 1986), which is accelerated by the lack of mixing of applied, acid-forming nitrogen fertilizers into the soil (Vitosh et al. 1985). This lack of mixing in no-tillage systems leads to a stratification of other fertilizer nutrients, in addition to nitrogen, within the profile and may inhibit root development and associated rhizosphere chemistry and biology.

Gaseous exchange of O_2/CO_2 between the plant root and the atmosphere may be the single most critical environmental factor controlling rhizosphere biology (Curl and Truelove 1986) that can be managed by soil tillage. Hamblin (1985) and Allmaras and Logsdon (1990) present extensive discussions of the close interactions of soil strength, bulk density, and water content in their effect on soil aeration. As summarized by Doran and Smith (1987), there is a greater preponderance of anaerobic organisms in the surface and, by implication, the rhizosphere, of no-tillage than in plowed soils. This difference was related to the more compact and wetter conditions of reduced tillage, but was very site specific.

SUMMARY

Soil organic matter can have both direct and indirect interactions and effects on rhizosphere ecology. Variations in amounts, composition, spatial distribution, and functions of organic matter in agricultural surface soils can greatly influence the physical, chemical, and biological environment. The dynamics of soil carbon and nitrogen cycles involve a wide variety of transformations, most of which include the soil organic matter fraction or its component humic substances and polysaccharides. Evidence of interactions at the root-soil interface confirms the dominant role played by microbes in the rhizosphere ecosystem, regarding both C-N biomass and biochemical transformations. Use of ^{14}C and ^{15}N isotopes has allowed researchers to follow photosynthates from shoots to roots and into the soil as rhizodeposits. The relationship between soil tillage and management of crop residues, as mediated through soil organic matter, is seen as a direct influence on plant root development and the rhizosphere environment.

An appreciation of carbon and nitrogen dynamics can be summarized by use of simple conceptual and mathematical models that simulate the effects of such factors as soil water content, root density, root exudation rates, and microbial growth constants on microbial and substrate concentrations in the rhizosphere.

REFERENCES

Aiken, G. R., D. M. McKnight, R. L. Wershaw, and P.
    MacCarthy, ed. 1985. Humic substances in soil,
    sediment, and water. New York: Wiley-Interscience.
Allison, F. E. 1973. Soil organic matter and its role in
    crop production. New York: Elsevier.
Allmaras, R. R., J. M. Kraft, and D. E. Miller. 1988.
    Effects of soil compaction and incorporated crop
    residue on root health. Annual Review of Phytopa-
    thology 26:219-243.
Allmaras, R. R., and S. D. Logsdon. 1990. Soil structural
    influences on the rhizosphere. In Rhizosphere dynam-
    ics, ed. J.E. Box, Jr. and L.C. Hammond, Boulder, CO:
    Westview Press.
Anderson, H. A., W. Bick, A. Hepburn, and M. Stewart.
    1989. Nitrogen in humic substances. In Humic sub-
    stances, vol. 2: In search of structure, ed. M. H. B.
    Hayes, P. MacCarthy, R. L. Malcolm, and R. S. Swift,
    223-253, Chichester, England: John Wiley and Sons.
Barber, D. A., and J. K. Martin. 1976. The release of
    organic substances by cereal roots into soil. New
    Phytologist 76:69-80.
Berg, P., and T. Rosswall. 1987. Seasonal variations in
    abundance and activity of nitrifiers in four arable
    cropping systems. Microbial Ecology 13:75-87.
Billes, G., and P. Bottner. 1981. Effets des racines
    vivantes sur la decomposition d'une litiere racinaire
    marquee au ^{14}C. Plant and Soil 62:193-208.
Biondoni, M., D. A. Klein, and E. F. Redente. 1988. Car-
    bon and nitrogen losses through root exudation by
    Agropyron cristatum, A. smithii and Bouteloua
    gracilis. Soil Biology and Biochemistry 20:477-482.
Blevins, R. L., W. W. Frye, and M. S. Smith. 1985. The
    effects of conservation tillage on soil properties.
    In A systems approach to conservation tillage, ed. F.
    M. D'Itri, 99-110. Chelsea, MI: Lewis Publishers.
Boller, B. C., and J. Nosberger. 1988. Influence of
    dissimilarities in temporal and spatial N-uptake
    patterns on ^{15}N-based estimates of fixation and
    transfer of N in ryegrass-clover mixture. Plant and
    Soil 112:167-175.
Boyd, S. A., and L. E. Sommers. 1990. Humic and fulvic
    acid fractions from sewage sludges and sludge-amended
    soils. In Humic substances in soil and crop sci-

ences, ed. P. MacCarthy, C. E. Clapp, P. R. Bloom, and R. L. Malcolm. Madison, WI: Soil Science Society of America.

Brady, N. C. 1984. The nature and properties of soils, 9th ed. New York: Macmillan.

Bremner, J. M., and R. D. Hauck. 1982. Advances in methodology for research in nitrogen transformations in soils. In Nitrogen in agricultural soils, ed. F. J. Stevenson, 467-502. Madison, WI: American Society of Agronomy.

Brophy, L. S., G. H. Heichel, and M. P. Russelle. 1987. Nitrogen transfer from forage legumes to grass in a systematic planting design. Crop Science 27:753-758.

Chen, Y., and T. Aviad. 1990. Effects of humic substances on plant growth. In Humic substances in soil and crop sciences, ed. P. MacCarthy, C. E. Clapp, P. R. Bloom, and R. L. Malcolm. Madison, WI: Soil Science Society of America.

Clapp, C. E., W. E. Larson, R. H. Dowdy, D. R. Linden, G. C. Marten, and D. R. Duncomb. 1984. Utilization of municipal sewage sludge and wastewater effluent on agricultural land in Minnesota. In Proceedings of the second international symposium on peat and organic matter in agriculture and horticulture, ed. K. Schallinger, 259-292. Bet Dagan, Israel: Volcani Center.

Clapp, C. E., S. A. Stark, D. E. Clay, and W. E. Larson. 1986. Sewage sludge organic matter and soil properties. In The role of organic matter in modern agriculture, ed. Y. Chen and Y. Avnimelech, 209-253. Dordrecht, Netherlands: Martinus Nijhoff.

Curl, E. A., and B. Truelove. 1986. The rhizosphere. Advanced Series in Agricultural Sciences, vol. 15. Berlin: Springer-Verlag.

Davenport, J. R., and R. L. Thomas. 1988. Carbon partitioning and rhizodeposition in corn and bromegrass. Canadian Journal of Soil Science 68:693-701.

Doran, J. W., and M. S. Smith. 1987. Organic matter management and utilization of soil and fertilizer nutrients. In Soil fertility and organic matter as critical components of production systems, ed. R. F. Follett, J. W. B. Stewart, and C. V. Cole, 53-72. SSSA Special Publication no. 19. Madison, WI: Soil Science Society of America.

Dowdy, R. H., A. M. Bidwell, D. R. Linden, and R. R. Allmaras. 1988. Corn root distributions as a function of tillage and residue management. In Proceedings of the Eleventh Conference of the International Soil Tillage Research Organization, 55-60. Edinburgh, Scotland.

Eaglesham, A. R. J., A. Ayanaba, A. Ranga Rao, and D. L. Eskew. 1981. Improving the nitrogen nutrition of maize by intercropping with cowpeas. Soil Biology and Biochemistry 13:169-171.

Ehlers, W. 1982. Penetrometer soil strength and root growth in tilled and untilled loess soil. In Proceedings of the Ninth Conference of the International Soil Tillage Research Organization, 458-463. Osijek, Yugoslavia.

Elliott, L. F., R. I. Papendick, and D. F. Bezdicek. 1987. Cropping practices using legumes with conservation tillage and soil benefits. In The role of legumes in conservation tillage systems, ed. J. F. Power, 81-89. Ankeny, IA: Soil Conservation Society of America.

Fisher, R. F., and E. L. Stone. 1969. Increased availability of nitrogen and phosphorus in the root zone of conifers. Soil Science Society of America Proceedings 33:955-961.

Follett, R. F., J. W. B. Stewart, and C. V. Cole, ed. 1987. Soil fertility and organic matter as critical components of production systems. SSSA Special Publication no. 19. Madison, WI: Soil Science Society of America.

Foster, R. C., A. D. Rovira, and T. W. Cock. 1983. Ultrastructure of the Root-Soil Interface. St. Paul, MN: American Phytopathology Society.

Frimmel, F. H., and R. F. Christman, ed. 1988. Humic substances and their role in the environment. Chichester, England: John Wiley and Sons.

Gallaher, R. N., and M. B. Ferrer. 1987. Effects of no-tillage vs. conservational tillage on soil organic matter and nitrogen contents. Communications in Soil Science and Plant Analysis 18:1061-1076.

Gieseking, J. E., ed. 1975. Soil components: Organic components, vol. 1. Berlin: Springer-Verlag.

Haller, T. N., and H. Stolp. 1985. Quantitative estimation of root exudation of maize plants. Plant and Soil 86:207-216.

Hamblin, A. P. 1985. The influence of soil structure on water movement, crop root growth, and water uptake. Advances in Agronomy 38:95-158.

Harley, J. L., and R. S. Russell, ed. 1979. The root-soil interface. London: Academic Press.

Hauck, R. D., and J. M. Bremner. 1976. Uses of tracers for soil and fertilizer nitrogen research. Advances in Agronomy 28:219-266.

Hayes, M. H. B. 1984. Structures of humic substances. In Organic matter and rice, 93-115. Los Banos, Philippines: International Rice Research Institute.

Hayes, M. H. B. 1985. Extraction of humic substances from soil. In Humic substances in soil, sediment, and water, ed. G. R. Aiken, D. M. McKnight, R. L. Wershaw, and P. MacCarthy, 329-362. New York: Wiley-Interscience.

Hayes, M. H. B., P. MacCarthy, R. L. Malcolm, and R. S. Swift, ed. 1989. Humic substances, vol. 2: In search of structure. Chichester, England: John Wiley and Sons.

Hayes, M. H. B., and R. S. Swift. 1978. The chemistry of soil organic colloids. In The chemistry of soil constituents, ed. D. J. Greenland and M. H. B. Hayes, 179-320. Chichester, England: John Wiley and Sons.

Helal, H. M., and D. R. Sauerbeck. 1983. Method to study turnover processes in soil layers of different proximity to roots. Soil Biology and Biochemistry 15:223-225.

Helal, H. M., and D. R. Sauerbeck. 1987. Direct and indirect influences of plant roots on organic matter and phosphorus turnover in soil. In Soil organic matter dynamics and soil productivity, ed. J. H. Cooley. The International Association for Ecology Bulletin 15:49-58.

Herman, M. 1985. Antagonistic activity of the rhizosphere mycoflora against Gaeumannomyces graminis under conventional and zero-tillage. Soil and Tillage Research 5:371-379.

Hunt, P. G. 1990. Microbial responses in the rhizosphere of agricultural plants. In Rhizosphere dynamics, ed. J.E. Box, Jr. and L.C. Hammond. Boulder, CO: Westview Press.

Jenkinson, D. S. 1977. Studies on the decomposition of plant material in soil. V. The effect of plant cover and soil type on the loss of carbon from ^{14}C-labelled ryegrass decomposing under field conditions. Journal of Soil Science 28:424-434.

Johnen, B. G., and D. R. Sauerbeck. 1977. A tracer technique for measuring growth, mass and microbial breakdown of plant roots during vegetation. _Ecological Bulletin_ (Stockholm) 25:366-373.

Kang, B. T., G. F. Wilson, and L. Spikens. 1981. Alley cropping maize (_Zea mays_ L.) and leucana (_Leucana leucocephala_ Lam.) in Southern Nigeria. _Plant and Soil_ 63:165-179.

Klemedtsson, L., P. Berg, M. Clarholm, J. Schnurer, and T. Rosswall. 1987. Microbial nitrogen transformations in the root environment of barley. _Soil Biology and Biochemistry_ 19:551-558.

Kononova, M. M. 1966. _Soil organic matter: Its nature, its role in soil formation and soil fertility._ 2nd English ed. Oxford, England: Pergamon Press.

Lal, R., D. DeVleeschauwer, and R. Molofa Nanje. 1980. Changes in properties of a newly cleared tropical Alfisol as affected by mulching. _Soil Science Society of America Journal_ 44:827-833.

Lambers, H. 1987. Growth, respiration, exudation and symbiotic associations: The fate of carbon translocated to the roots. In _Root development and functions_, ed. P. J. Gregory, J. V. Lake, and D. A. Rose, 125-145. Cambridge, England: Cambridge University Press.

Legg, J. O., F. W. Chichester, G. Stanford, and W. H. DeMar. 1971. Incorporation of ^{15}N-tagged mineral nitrogen into stable forms of soil organic nitrogen. _Soil Science Society of America Proceedings_ 35:273-276.

Lory, J. A., M. P. Russelle, and G. H. Heichel. 1988. Quantification of symbiotically fixed nitrogen deposited in soil surrounding alfalfa roots and nodules. p. 241. _Agronomy Abstracts._ Madison, WI: American Society of Agronomy.

MacCarthy, P., C. E. Clapp, P. R. Bloom, and R. L. Malcolm, ed. 1990. _Humic substances in soil and crop sciences: Selected readings._ Madison, WI: Soil Science Society of America.

Martin, J. K. 1977. Factors influencing the loss of organic carbon from wheat roots. _Soil Biology and Biochemistry_ 9:1-7.

Meharg, A. A., and K. Killham. 1988. A comparison of carbon flow from pre-labelled and pulse-labelled plants. _Plant and Soil_ 112:225-231.

Milchunas, D. G., W. K. Lauenroth, W. K. Singh, C. V. Cole, and H. W. Hunt. 1985. Root turnover and production by ^{14}C dilution: Implications of carbon partition in plants. _Plant and Soil_ 80:353-365.

Newman, E. I., and A. Watson. 1977. Microbial abundance in the rhizosphere: A computer model. Plant and Soil 48:17-56.

Ofori, F., J. S. Pate, and W. R. Stern. 1987. Evaluation of N^2-fixation and nitrogen economy in a maize/cowpea intercrop system using ^{15}N dilution methods. Plant and Soil 102:149-160.

Orlov, D. S. 1985. Humus acids of soils. New Delhi: Amerind Publishing Co.

Page, A. L., T. L. Gleason, J. E. Smith, I. K. Iskander, and L. E. Sommers, ed. 1983. Utilization of municipal wastewater and sludge on land. Riverside, CA: University of California.

Papastylianou, I. 1988. The ^{15}N methodology in estimating N^2 fixation by vetch and pea grown in pure stand or mixture with oat. Plant and Soil 107:183-188.

Patra, D. D., M. S. Sachdev, and B. V. Subbiah. 1986. ^{15}N studies on the transfer of legume-fixed nitrogen to associated cereals in intercropping systems. Biology and Fertility of Soils 2:165-171.

Paul, E. A., and R. M. N. Kucey. 1981. Carbon flow in plant microbial associations. Science 213:373-374.

Poth, M., J. S. Favre, and D. D. Focht. 1986. Quantification by direct ^{15}N dilution of fixed N_2 incorporation into soil by Cajanus cajan (pigeon pea). Soil Biology and Biochemistry 18:125-127.

Reid, J. B., and M. J. Goss. 1983. Growing crops and transformations of ^{14}C-labelled soil organic matter. Soil Biology and Biochemistry 15:687-691.

Rice, E. L., and S. K. Pancholy. 1972. Inhibition of nitrification by climax vegetation. American Journal of Botany 59:1033-1040.

Rovira, A. D. 1969. Plant root exudates. Botanical Review 35:35-57.

Rovira, A. D., and C. B. Davey. 1974. Biology of the rhizosphere. In The plant root and its environment, ed. E. W. Carson, 153-204. Charlottesville, VA: University of Virginia Press.

Rovira, A. D., R. C. Foster, and J. K. Martin. 1979. Origin, nature and nomenclature of the organic materials in the rhizosphere. In The soil-root interface, ed. J. L. Harley and R. Scott-Russel, 1-4. London: Academic Press.

Rovira, A. D., and B. M. McDougall. 1967. Microbiological and biochemical aspects of the rhizosphere. In Soil biochemistry, ed. A. D. McLaren and G. H. Peterson, 417-463. New York: Marcel Dekker.

Sangina, N., K. Mulongoy, and A. Ayanaba. 1988. Nitrogen contribution of Leucaena/Rhizobium symbiosis to soil and subsequent maize crop. Plant and Soil 112:137-141.

Sauerbeck, D. R., and B. G. Johnen. 1977. Root formation and decomposition during plant growth. In Proceedings of the international symposium on organic matter studies, 141-148. Vienna, Austria: International Atomic Energy Agency.

Schnitzer, M. 1985. Nature of nitrogen in humic substances. In Humic substances in soil, sediment, and water, ed. G. R. Aiken, D. M. McKnight, R. L. Wershaw, and P. MacCarthy, 303-325. New York: Wiley-Interscience.

Schnitzer, M. 1990. Selected methods for the characterization of soil humic substances. In Humic substances in soil and crop sciences, ed. P. MacCarthy, C. E. Clapp, P. R. Bloom, and R. L. Malcolm. Madison, WI: Soil Science Society of America.

Schnitzer, M., and S. U. Khan, ed. 1978. Soil organic matter. New York: Elsevier.

Smith, M. S., and J. M. Tiedje. 1979. The effect of roots on soil denitrification. Soil Science Society of America Journal 43:951-955.

Sparling, G. P., M. V. Cheshire, and C. M. Mundie. 1982. Effect of barley plants on the decomposition of ^{14}C-labelled soil organic matter. Journal of Soil Science 33:89-100.

Stevenson, F. J. 1982. Humus chemistry: Genesis, composition, reactions. New York: Wiley-Interscience.

Stevenson, F. J. 1986. Cycles of soil: Carbon, nitrogen phosphorus, sulfur, micronutrients. New York: Wiley-Interscience.

Stevenson, F. J., and X. T. He. 1990. Nitrogen in humic substances related to soil fertility. In Humic substances in soil and crop sciences, ed. P. MacCarthy, C. E. Clapp, P. R. Bloom, and R. L. Malcolm. Madison, WI: Soil Science Society of America.

Swift, R. S. 1985. Fractionation of soil humic substances. In Humic substances in soil, sediment, and water, ed. G. R. Aiken, D. M. McKnight, R. L. Wershaw, and P. MacCarthy, 387-408. New York: Wiley-Interscience.

Ta, T. C., M. A. Faris, and F. D. H. Macdowall. 1989. Evaluation of ^{15}N methods to measure nitrogen transfer from alfalfa to companion timothy. Plant and Soil 112:137-141.

Thomas, G. W. 1986. Mineral nutrition and fertilizer placement. In No-tillage and surface-tillage agriculture: The tillage revolution, ed. M. A. Sprague and G. B. Triplett, 93-116. New York: John Wiley and Sons.

Tiessen, H., and J. W. B. Stewart. 1983. Particle-size fractions and their use in studies of organic matter: II. Cultivation effects on organic matter composition in size fractions. Soil Science Society of America Journal 47:509-514.

Todd, R. L., and J. E. Giddens, eds. 1984. Microbial-plant interactions. ASA Special Publication no. 47. Madison, WI: American Society of Agronomy.

Trofymow, J. A., D. C. Coleman, and C. Cambardella. 1987. Rate of rhizodeposition and ammonium depletion in the rhizosphere of axenic oat roots. Plant and Soil 97:333-344.

Vitosh, M. L., W. H. Darlington, C. W. Rice, and D. R. Christenson. 1985. Fertilizer management for conservation tillage. In A systems approach to conservation tillage, ed. F. M. D'Itri, 89-98. Chelsea, MI: Lewis Publishers.

Waksman, S. A. 1938. Humus: Origin, chemical composition and importance in nature. Baltimore: Williams and Wilkins.

Whipps, J. M. 1984. Environmental factors affecting the loss of carbon from the roots of wheat and barley seedlings. Journal of Experimental Botany 35:767-773.

Woldendorp, J. W. 1975. Nitrification and denitrification in the rhizosphere. In La Rhizosphere, ed. F. Mangenot, 89-107. Societe Botanique Francaise Colloquium Rhizosphere. Paris, France: Societe Botanique Francaise.

Wood, M. 1987. Predicted microbial biomass in the rhizosphere of barley in the field. Plant and Soil 97:303-314.

Dan R. Upchurch
Howard M. Taylor

4   Tools for Studying Rhizosphere Dynamics

Roots perform at least three functions required for plant growth. They anchor the plant to the soil and support the base of the stem, much as the foundation of a building anchors and supports the building. Also, the root system is the primary organ for supplying nutrients and water to the plant. Carbon is the only element used extensively by plants that is not taken up primarily through the root system. Third, plant root systems synthesize or mediate the action of various growth regulating compounds, which include but are not limited to the kinins and ethylene.

The environment of both the shoot and the root affect the development of the root system. Soil micro-flora and fauna have a direct impact on the root system. Microbes thrive in the nutrient-rich rhizosphere, using organic material exuded by the root as an energy source. Some microbes release nutrients and growth-regulating compounds used by the plant. These same microbes can also attack living root tissue.

Like the shoot, roots are sensitive to both high and low temperature extremes. The inclination angle of root growth with respect to gravity is influenced by the temperature of the soil surrounding the root (Kaspar et al. 1981; Mosher and Miller 1972). Low temperatures can reduce root growth (Stone and Taylor 1983) and impede nutrient and water uptake through a reduced rate of metabolism. High temperatures are lethal to roots, which are not protected by the cooling action of evaporating water as are shoots.

Water in the soil surrounding roots is a solution containing various dissolved salts. This solution is the medium that provides the nutrients required for plant growth. Uptake of nutrients by the root is not an entirely passive, mass-flow process. Some ions in the soil solution are extracted preferentially by the root, while others are excluded from the uptake stream. Additionally, the soil solution can contain toxic compounds or toxic concentrations of required nutrients. High concentrations of ions in the solution can also result in a significant osmotic potential gradient from the root toward the soil solution. This can impede nutrient and water uptake across the endodermis of the root.

Certain elements are required by the plant for root growth to occur. For example, root elongation will cease if calcium and boron are absent from the soil solution (Loneragan 1979). The soil solution may also act as the transport agent for various organic molecules that impact plant performance. Some plant residues contain substances that may be either detrimental or beneficial to the performance of other plants.

Another factor, which may cause the greatest response in the root system, is the environmental evaporative demand. High evaporative demand results in a shift in the root-shoot ratio caused by a larger fraction of the photosynthate being partitioned to the root system (Malik et al. 1979). This reallocation is apparently an adaptive response to increase the capacity of the soil-plant system to supply water to the shoot.

The study of any biological organism is difficult because of the complex interactions that exist. The environment in which the root system exists compounds this difficulty. The root system is shielded from direct observation by the soil matrix. Most measurements of the root system are disruptive, if not totally destructive, to the root system and its environment. Manipulation of the root environment to allow observation of the root system always results in some question about the effect of the measurement on the observation. Sometimes the results are an artifact of the measurement technique.

A further complication involved in studying roots arises from the spatial variability of roots and soil properties. Roots are not uniformly distributed in the soil with depth or lateral distance from the crown of the plant. Soil properties are not uniform at the scale of an individual root; therefore, we should not expect the root system to function uniformly throughout.

The root system and its environment are dynamic. New roots are constantly being produced, while others die. The root environment changes as the rhizosphere is dried and depleted of nutrients through each day by plant extraction

and is then re-wetted during the night by redistribution
when transpiration is minimal (Baker and van Bavel 1986).
As root systems extract water from the soil, a water potential gradient develops around the root. If plant transpiration exceeds the rate of water movement through the soil,
the soil water content and potential around the root will
be reduced. Reduced soil water content and potential can
result in desiccation of the root tissue if the root is not
protected through suberization. To elongate, roots must
overcome a larger resistance when the soil water content is
reduced. Over longer periods of time the soil will become
progressively drier when rain and irrigation do not keep
pace with evapotranspiration. The temperature near the
soil surface may vary over a 30°C range during a single
day. In comparison, the temperature at greater depths is
stable but may exhibit a change sufficient to impact root
growth and function when the entire growing season is considered (Mason et al. 1982; Stone et al. 1983).

Changes in the size of a root system through the life
of the plant are tremendous. Initially, a single root
emerges from the seed and functions as the only supplier of
water and nutrients to the plant. Ultimately, the total
length of roots associated with an individual plant may be
several kilometers. As the plant shoot grows, the root
system grows and branches into a complex system of inter-
connected conduits for transporting water and nutrients
from the soil to the shoot. The root system explores new
soil as it grows, providing an increased supply of nutrients and water to the shoot. When the plant changes from
the vegetative phase into the reproductive phase, the sink
demand of developing seeds reduces the supply of carbohydrates to roots. This generally results in a net loss of
small roots. Even during this phase some new roots will be
produced, but at a rate less than the rate at which roots
are lost.

The study of rhizosphere dynamics involves many disciplines, each requiring specialized research procedures. We
will discuss the procedures used in quantifying the total
amount of roots and their distribution in the growing
medium. These procedures can be divided into four broad
categories:

1) destructive, field-oriented techniques;
2) nondestructive, field-oriented techniques;
3) destructive, laboratory-oriented techniques;
4) and nondestructive, laboratory-oriented techniques.

Within each of these categories, there may be several appropriate procedures. The choice of a particular procedure depends upon the research question and the available resources. The procedure to be used in a project may be dictated by the root parameter of interest. For example, in an experiment in which the total root dry weight at the end of the vegetative phase is needed, a nondestructive, field-oriented technique would not be particularly useful; whereas, a destructive, laboratory technique would be well suited to measuring this parameter.

There is a significant amount of literature describing the details of various techniques for studying roots. Bohm (1979) reviewed this literature extensively, and his book serves as a major resource for descriptions of procedures that are currently available.

DESTRUCTIVE TECHNIQUES

Destructive techniques were the earliest procedures developed for root studies. They are generally labor-intensive but require a minimal amount of equipment. The root system and its environment are destroyed during the operation, precluding the possibility of repeated observations on the same root system. These techniques are applied for point-in-time quantification of the root system. To study the dynamics of root system development using these tools, observations must be made at a new site for each date. This new site requirement introduces errors because of soil variability and nonuniform distribution of roots within the soil. Root distribution at a particular point in time and space is most accurately determined by a destructive technique. Destructive techniques are therefore used in calibration of new techniques. Results in the literature concerning the variability of the destructive techniques suggest that "big is best." Increasing sample size results in decreased sampling variability. This reduction in sample variability is attributable to inclusion of the natural spatial variability in individual samples.

Destructive techniques also require the most time to apply. Both collecting and processing the samples require large inputs of time and labor. This time requirement may exceed ten person hours per sample for large soil monoliths.

Field-oriented Destructive Techniques

It is assumed that a field situation provides the most realistic root growth data. Root systems in a field will develop under "natural" conditions: ambient climate and soil conditions affect their growth and development. Unfortunately, the variability of climate and soil properties precludes the use of the field environment in many research projects. It is difficult, if not impossible, to define a normal environment for a location. Total rainfall and rainfall distribution will vary between years, as will most other climatic factors. Also, soil properties vary spatially and temporally. This variability makes interpretation of root data from field experiments difficult. The root system observed may be a result of the treatments imposed or of environmental conditions during the experiment.

Excavation was the earliest procedure allowing inspection of the root system (Hales, 1727). Virtually every technique used for destructive observation of root systems in either a field or a controlled environment is a variation of the excavation technique. The basic process of carefully removing the soil from around the roots, however, has remained the same. Weaver (1926) presented details of the excavation process, establishing it as an appropriate scientific investigative procedure.

Excavation. In its classical use, the excavation technique allows the determination of the relative position of individual roots. In practice, a trench is opened some distance from the base of the plant. This distance should be sufficient to allow location of the longest lateral root, but short enough to minimize the volume of soil removed. Soil removal begins at the wall of the trench, and the roots are carefully exposed. Exposed roots should be held in position as soil removal proceeds. The entire root system can then be collected or recorded with drawings or photographs. Improvements in this technique have come primarily from the procedures used in removing the soil. Fine roots are fragile and difficult to isolate from the soil. Removing the soil with small picks and brushes tends to damage a large fraction of these fine roots. Water pressure and pressurized air have been used to remove soil while preserving fine roots. Observation of the entire root system of a field-grown plant requires an excavation technique. Excavation techniques are also well adapted to ecological studies in which small details of the root system may be highly significant. Examples are the response of individual roots to a high-strength soil layer, or to an acid subsoil. Figure 4.1 shows the root system of a cotton plant that was excavated from the upper 200 mm of the soil profile. This type of excavation allows measurement of morphological characteristics of the root system such as branching intensity and root diameters. In some soils, excavation is the only option available because of soil characteristics. A stony soil may preclude the use of soil cores or other techniques because of the interference of the stones with sample collection. Excavation procedures are not well suited to statistical analysis because of the time required to obtain multiple samples.

Monoliths. This variation of the excavation technique involves collecting blocks of soil (monoliths) containing roots, transporting the monoliths to a washing facility, and separating the roots from the soil. Monoliths are used in place of simple excavation when root system parameters must be quantified. After the roots are separated from the soil, they can be quantified by any of several methods. Various parameters, such as root length, root surface area, or root dry weight, can be measured after the separation procedure.

The monolith technique has been used with many variations in the size and shape of monoliths, and in collection procedures. Generally, one vertical wall of a trench is smoothed by careful removal of soil. The wall is then inscribed, defining the size of each soil block. Alternately, a free-standing block of soil may be encased with some solid material. The size of the soil blocks depends on the research project and resources available for moving large soil blocks. Typical soil block volumes are one to five liters. Larger monoliths have been collected, however. The framed monolith technique of Nelson and Allmaras (1969) is an example of an application of this technique in which a large soil block was removed (1.5 x 0.4 x 1.0 m). (See Figure 4.2.)

Schuurman and Goedewaagen (1965) described a monolith technique in which the root system is held in its natural position as the soil is removed. The pin or needle-board method was originally described by Goff (1897). In this technique, a monolith is obtained with a standard approach, but pins are inserted through the soil at regular spacings prior to washing. As the soil is washed away, the roots are held in their approximate natural position by the pins. Generally, one wall of a trench is smoothed, a rigid board is placed against the wall, and pins are driven through the board into the soil. The monolith is then removed with the pins in place. Knitting needles, nails, stiff wire cut to

Figure 4.1. Cotton root system excavated from the top 200 mm of the soil profile.

Figure 4.2. Framed monolith being extracted using the procedure of Nelson and Allmaras (1969).

the appropriate length, and bicycle spokes have been used as pins. Boards with holes pre-drilled to establish the pattern for pin insertion and then backed with a stronger material after pin insertion, are also used.

The needle-board method allows both qualitative observation of the relative position of roots and quantitative determination of various root parameters. Roots can be photographed in their natural position, and overall root system morphology can be observed. Removal of the roots from the pinboard allows for quantitative measurements.

The entire root system normally is not collected when using this method. Only those roots that grow entirely within the soil volume of the monolith are retained during washing. Roots penetrating the faces of the monolith normally are lost during washing. The resources available for lifting the soil block and the length of pins that can be inserted limit the size of monoliths. Generally, 1.0 m depths and widths with a thickness between 0.1 and 0.5 m is the limit for monolith size. The framed monolith technique of Nelson and Allmaras (1969) is again an example of larger monolith sizes.

Trench profiles. A somewhat less destructive and less labor demanding technique for observing roots in a thin layer of soil at the wall of a trench was first used by Weaver (1919). This procedure involves opening a trench and smoothing one wall such that careful soil removal will expose the roots. The thickness of the layer of soil removed may be between 10 and 100 mm. The length of exposed roots can then be measured, individual roots counted, or the rooting pattern drawn on an overlay. A grid may be placed over the wall prior to counting the roots to facilitate recording their positions. When important variations in soil properties are visible on the wall, these can be outlined on the overlay to aid in interpreting the root distribution.

For quantifying the rooting patterns, a thin layer of soil is preferred. Removal of thick layers of soil may result in the exposed roots becoming tangled and moving from their original position, causing results to be more qualitative. The number of roots exposed can be used to estimate root length density by assuming a specific length associated with each root. Estimated total length is then divided by the volume of soil removed in exposing the roots. Often the length of each root is assumed to be the thickness of the soil layer removed.

There are few limitations to depth and width of the trench wall exposed for measurements. Because all measurements are accomplished in the field, the primary limitation is time. Roots dry quickly when exposed to the atmosphere. Dried roots are fragile and are easily lost during counting. The amount of time required will depend on the procedure used for counting and on the density of roots at the trench wall.

An example of this procedure is shown in Figure 4.3, a photograph of a profile wall with a grid in place. Soil was removed from each grid square beginning at the bottom and proceeding upward.

Soil cores. Here we will discuss techniques in which comparatively small soil volumes are collected. The samples may be intact soil cores or volumes of disturbed soil. This technique is similar in principle to the monolith methods because soil samples of known volumes are collected, and roots are separated from the soil and then quantified. The primary difference is in the size of the soil volume collected. Because only a small fraction of the root system is collected using this method, length or dry weight of the entire root system can only be estimated. Usually, the sample volume is less than one liter. The procedures for collecting the samples range from auguring with hand-operated tools to collecting intact cores with hydraulically operated equipment. The underlying assumption is that the distribution of roots can be described by observing the density of roots in a small fraction of the root zone. It is assumed that the roots are uniformly distributed horizontally or that an adequate number of samples can be collected to characterize the heterogeneity of root distribution. In studies of row crop rooting patterns, the horizontal distribution can exhibit extreme anisotropy, and a sampling scheme must account for this distribution. Often samples are taken at specific distance intervals away from the plant row, with sufficient replication of locations to account for variation along the row. Many systems for collecting cores are described in the literature. Generally, soil is removed from a cylindrical cavity to the depth of interest and divided into subsamples by depth, often at 0.1 to 0.2 m intervals. A soil core 76 mm in diameter is shown in Figure 4.4. This core was collected using a hydraulically driven steel tube designed to collect an intact core. The core is slightly smaller in diameter than the sampling tube, minimizing compaction of the sample. If the depth of interest is less than one meter, the samples can be collected using hand-operated equipment (Foale and Upchurch 1982); however, powered equipment is generally desirable. The diameter of the soil core is between 30 and

Figure 4.3.   Trench profile wall with grid in place. Each grid square is 50 x 50 mm.

Figure 4.4. Intact soil core, 76 mm in diameter, collected with a steel sampling tube hydraulically driven into the soil.

100 mm, with a 50 mm diameter being the most commonly reported size.

The number of cores collected depends on the resources available, the detectable difference in rooting desired, and core size. Because cores are smaller in volume than monoliths, a larger number of samples must be collected. A minimum of three cores per plot should be collected so that a statistical confidence can be computed. As the magnitude of the desired detectable difference decreases, the number of samples required increases exponentially.

A variation, called the core-break technique, involves determining the number of roots exposed on the two faces of a broken soil core. Drew and Saker (1980) developed a theoretical conversion of the number of roots exposed at the break to the root length density in the core. The time required to use the core-break technique is much less than that required for washing roots free from soil and quantifying the total root length. The core-break technique is difficult to use when the soil is dry, crumbles easily, or is stony.

A reduction of time and labor is the primary advantage of soil-coring techniques over the monolith and trench techniques. Because a relatively large number of cores can be collected in a short time without severe damage to the plot, the dynamics of the root system may be statistically quantified. Soil coring is less destructive to the plant root system and total soil volume than the monolith and trench techniques. Although the collected sample is destroyed during the measurement, the surrounding soil is left intact and the plant only slightly disturbed. The results obtained from soil coring have a large variance, which limits their usefulness in detecting small differences. The number of samples required may become unreasonably large, especially when the rooting density in the sampled zone is low. Soil cores are difficult to obtain from stony soils or dry soils that crumble easily. The volume of the sample is often difficult to determine because of soil compaction during coring or loss of part of the sample when the soil is dry. Soil coring is not adapted to studies of root morphology because only a small portion of the root system is collected. Even with these limitations soil coring is still the most frequently used procedure for studies of root systems and is often the calibration standard for other techniques.

Laboratory-oriented Destructive Techniques

These techniques are those traditionally used in a laboratory, growth chamber, or greenhouse situation. They are advantageous because researchers are able to control factors that are not controllable in the field. The specific sampling techniques are similar to those used in the field, however. Often the plants are grown in pots containing processed soil. The processing results in destruction of the natural soil structure, including breaking up of aggregates and possibly the rearrangement of primary and secondary soil particles. Required equipment is minimal and labor costs are high. A major advantage is that collection of the entire root system is possible. Of course, root growth is affected by the pot walls, thereby affecting the root system size.

The procedures described in the previous section, with minor alterations, can be applied to root systems grown in controlled environments. The most common procedure used in controlled environments is a hybrid of the monolith and excavation techniques. Washing separates the entire root system from the soil contained in a pot. To determine the root distribution with depth, the pot is divided into sections. Alternately, the three-dimensional configuration of the root system can be determined by holding the root system in place with pins or wire during the washing process. The dimensions of the container in which the plant is grown can be selected for a specific research hypothesis. For example, tall containers with a small diameter are appropriate if the rooting depth is of interest. A split root system may be used to impose treatments on different sections of a root system (Adeoye and Rawlins, 1981; Anghinoni and Barber, 1980; Emanuelsson, 1984; Lonkerd and Ritchie, 1979).

Washing Procedures

A prerequisite of the destructive techniques is the separation of roots from the soil. Bohm (1979) describes several procedures for this purpose. Most of them involve applying water pressure or soaking in water and then screening the supernatant to separate the roots from the soil and suspension.

Soil samples are often stored for some time between collection and washing. Care must be taken to preserve the roots during this time. Several processes are available for preserving roots, such as storing the sample in alcohol, formalin, or some other preservative or drying the sample. The samples are often refrigerated or frozen and then thawed before washing.

The washing process involves dispersing the soil so the roots are free from clinging soil particles, and then separating the roots by passing the suspension through a sieve. The soil can be dispersed by adding a chemical agent such as hexametaphosphate to the solution or simply by agitating the solution. If the soil contains carbonates, dilute acid added to the sample will cause the formation of carbon dioxide gas, which will aid in dispersing the soil. The simplest and most economical method is to use water pressure to wash the soil from the roots. Samples are placed on or poured over a fine-mesh sieve, and water under pressure washes the soil through the sieve while roots remain on the sieve. There have been reports in the literature on various machines for washing roots from soil samples. There is presently at least one machine commercially available (Smucker et al. 1982). This machine, manufactured by Gillison Manufacturing[1], uses a combination of water pressure and flotation to separate roots from soil.

NONDESTRUCTIVE TECHNIQUES

Nondestructive root studies generally require a larger input of equipment than of labor; however, the labor requirement remains substantial. These techniques are often less precise than the destructive methods but allow repeated measurements of one root system or location, which facilitates studying root system dynamics. Interest in these techniques has increased in recent years because of their reduced labor requirements. Certain of these techniques were developed to provide answers to specific questions rather than to replace destructive techniques, while others were developed primarily as replacements for more destructive techniques.

---

[1]Mention of a trademark or proprietary product in this essay does not constitute a guarantee or warranty of the product by the United States Department of Agriculture or Texas Tech University, and does not imply their approval to the exclusion of other products that may also be suitable.

Field-oriented Nondestructive Techniques

The rationale for the use of nondestructive field-oriented techniques for observing roots is the same as that for the destructive techniques: the field environment provides the most realistic, but most difficult, situation for studying roots. Generally, these techniques were developed with the intention of replacing the more labor-intensive, destructive techniques. Massive excavation of root systems requires the availability of human resources simply because of the amount of soil that must be moved. Increasing labor costs have made excavation techniques uneconomical and have provided the incentive for developing the nondestructive tools. Recent availability of various new technologies has also acted as a stimulus for developing some of the nondestructive techniques. These techniques were originally described in the literature prior to development of the technology required to provide a truly useful tool.

Rhizotrons. One of the earliest nondestructive techniques for observing root growth is the rhizotron, a root observation laboratory (Rogers 1969). A rhizotron can be considered either as a field or laboratory oriented facility, depending upon the perspective of the researcher. Often plants are grown in disturbed soil, which is characteristic of laboratory studies. The shoots exist in a field environment, however, and therefore are subject to variations in the weather.

Rhizotrons are covered underground cellars or walkways with clear windows on one or both sides. The windows may be in contact with the natural soil surrounding the cellar or may be one wall of a soil-filled box. A typical rhizotron installation is shown in Figure 4.5. Plants are grown in the soil adjacent to the window such that some of the roots intersect the window and can be observed. Rhizotron designs have been published by Fordham (1972), Freeman and Smart (1976), Glover (1967), Hilton et al. (1969), Huxley and Turk (1967), Rogers (1969), Soileau et al. (1974), Taylor (1969), and Taylor and Willatt (1981). Rhizotron construction techniques, their operation, and the types of experiments being conducted in them have been reviewed by Huck and Taylor (1982) and Taylor and Willatt (1981).

Rhizotrons have several advantages over other root study methods. Successive measurements are made on the same plants and estimates of relative root growth are made rapidly. Instruments and sensors can be installed easily in rhizotrons to measure soil conditions such as temperature or soil water potential.

Figure 4.5. Typical rhizotron facility, showing the underground walkway lined with clear windows through which roots are observed.

Rhizotrons also have several disadvantages, the primary one being cost. The rhizotron at Auburn, Alabama, cost approximately $40,000 when constructed in 1969; today a similar facility costs approximately $100,000. The rhizotron located at Ames, Iowa, cost about $20,000 in 1973. For most rhizotrons, the aerial environment in which the shoots are grown is affected by the presence of various structures surrounding the site and is therefore likely to be different from that of the nearby fields. The covered underground walkway can have only a shallow soil layer over it and will therefore require a modified irrigation regime to support plant growth. If it is not covered with soil and cultivated, it may induce a significant level of advective energy transfer to the plants grown in the rhizotron. Rhizotrons require continual maintenance but are available for many experiments; hence, they are cost effective if a large number of scientists are located near the facility. Because of the high maintenance cost it is appropriate to have multiple projects and multiple investigators using the facility continuously.

Minirhizotrons. Originally proposed by Bates (1937), minirhizotrons were not widely used until recently because of technological limitations. This technique is a variation of the more traditional rhizotrons in that roots are observed growing in soil behind a clear material. Minirhizotron were devised and are being used primarily to replace soil core sampling in the field. The minirhizotron system is almost exclusively a field-oriented technique, although there has been at least one application in a greenouse (Rush et al. 1984).

This technique involves installing a clear tube in the soil and lowering a device into the tube that will allow roots intersecting the tube to be seen with the naked eye, photographed, displayed on a television monitor or recorded on video tape. Tubes are commonly installed a few days after planting or emergence. They are installed at angles of thirty to forty-five degrees from vertical to prevent roots from following the tube, which would distort the results. With these angles, roots do not often follow the upper side of the tube for extended distances. They may follow the underside, however, if there is not intimate contact between the tube and the soil. Care must be taken to establish contact between the soil and the tube while not compacting the soil. Markings are placed on the tube prior to installation so that permanent reference points can be located during observation. These are often a series of horizontal marks inscribed on the tube, with the accompanying depth intervals printed on the tube.

Bates (1937) originally used a mirror and battery-operated lamp mounted on the end of a stick to see roots intersecting a tube. The mirror-and-stick method of observation limited the usefulness of this technique because roots at depths below about one meter could not be seen without the aid of a telescope. With the application of new technologies to the observation procedure, results improved. Waddington (1971) replaced the mirror-and-stick arrangement with a coherent fiber-optic scope and right-angle viewing attachment, which improved the quality of the root image. Sanders and Brown (1978) included a 35mm camera with a fiber-optic duodenoscope to further improve images of the intersecting roots. Dyer and Brown (1980) replaced the 35mm camera with a black-and-white video camera. Upchurch and Ritchie (1984) described a system involving a color video camera with a right-angle viewing head. While the camera was lowered into a tube, a picture of the intersecting roots was recorded on video tape. This battery-operated system was mounted on a back-pack which could be carried in the field (Figure 4.6). This technology has improved the quality of the image such that any root intersecting the tube can be seen.

Various techniques are available for quantifying the intersections of roots with minirhizotron tubes. Bates (1937) drew pictures of the roots that intersected the tube. Later, Gregory (1979) measured the length of roots in contact with the tube. This length was used for relative comparisons between depths or treatments or converted to an estimate of root length density. The conversion to root length density involves one of several assumptions concerning the volume of soil to associate with the measured root length. It must be assumed that the observed roots are all contained within a soil cylinder of known volume, all roots within this volume can be seen, and that the density of roots in this layer of soil is representative of the density in the bulk soil. Upchurch and Ritchie (1983) proposed that the number of roots visible at the soil/tube interface, rather than their length, was the appropriate parameter. Upchurch (1985) developed a theoretical relationship between the number of intersecting roots and the root length density in the soil surrounding the minirhizotron. The results from this relationship adequately predicted the measured root length density.

One of the greatest limitations of the minirhizotron system is the number of tubes required to accurately estimate rooting. Upchurch and Ritchie (1983, 1984) and Up-

Figure 4.6. Minirhizotron system being used in a cotton field. All required electronic equipment is mounted on the backpack. A color video camera is being lowered into a 57 mm diameter tube buried in the soil, and the image of intersecting roots observed on a monitor.

church (1985) reported observations that suggest that a minimum of eight tubes may be required for the mean value to correlate with root length density of a single plot. This number is not substantially different, however, from the number of soil cores required for the same estimate. Although no explanation has been provided, near-surface (upper 0.2 to 0.4 m) estimates of rooting from this system are consistently low, when compared to estimates from soil cores. Variability in the results is high at or near the maximum rooting depth, where rooting density is low. Data from soil cores exhibit similar patterns of variability.

The minirhizotron system provides the opportunity for repeated observations of the sample location, eliminating one source of error. Although it requires thirty to forty-five minutes to install each minirhizotron, routine observation of roots in each tube requires about ten to fifteen minutes by two experienced individuals. At present, the equipment needed to observe and record color video pictures of roots is commercially available at a cost of about $15,000 (Circon Corporation, Santa Barbara, California).

Several extensions to the use of minirhizotrons have been reported or are under development. The minirhizotron system has been used to study root pathogens (Rush et al. 1984). Upchurch, McMichael, and Taylor (1988) described a method using the ratio of top to bottom root intersections to determine the dominant orientation of the root system. Procedures are also being developed to use this system in various studies of rhizosphere flora and fauna, especially in ecological studies. A complete review of this technique is available in a symposium publication (Taylor 1987).

Tracers. This group of procedures includes those techniques that might also be called indirect procedures. The uptake or translocation of some material, the tracer, implies the activity or location of the roots. The tracer may be a radioactive isotope, nonradioactive isotope, a stain, or even water.

Water extraction from the soil profile is the simplest of these tools. The change in water content at a given depth indicates the presence and activity of roots at that depth. When a plant is growing entirely on stored soil water, this can be a very efficient procedure for describing the relative distribution of root activity (Stone et al. 1976). Replacement of soil water by rainfall or irrigation, however, makes interpretation of the results difficult (Bohm et al. 1977). For example, water can be extracted from a given depth without a noticeable change in water content. Upward or downward flow of water in response to a hydraulic gradient can mask uptake from a layer. Redistribution of water can occur for several days after initial input into the soil. A change in water content can also occur in soil below the depth of rooting. This occurs when the uptake of water at a given depth creates a gradient in water potential sufficient to cause upward flow of water from lower soil layers. Drainage of water out of the root zone may occur and be confused with water uptake by the root system. Finally, roots can also be present in a layer, but not extract water because the soil layer is too dry.

Arrival of roots at a specific location in a soil profile can be determined by injecting a tracer at that location and then detecting that tracer in the shoot. The tracer is assumed to be immobile after injection, and uptake of the tracer is assumed to occur upon arrival of the roots at that location (Pearson 1974). Many different tracers have been used in this type of work. The tracer may be a radioactive material that can be detected in the shoot with a scintillation or Geiger counter. Atomic absorption techniques can detect a stable isotope used as a tracer. Plant toxins have been used as tracers (Robertson et al. 1985) so that the plants show visible stress or are killed when the roots intersect the injection zone.

Tracers can also be injected in the shoot and their presences detected in the soil (Racz et al. 1964). After injection into the shoot, the tracer is allowed to redistribute in the plant before collection of soil samples. Presence of the tracer in a soil sample indicates the presence of roots. It must be assumed that the tracer is translocated to all roots of interest and the tracer activity in the soil is related to the amount or activity of root material. Roots labeled with a radioactive tracer can be detected in situ by placing auto-radiographic film in the soil or on a smoothed soil wall.

Capacitance or resistance. Electrical capacitance has been suggested by Chloupek (1972) as a technique for estimating the extent of a root system. An electrode is connected to the base of a plant and another is inserted in the soil some distance from the base of the plant. The electrical capacitance of this system indicates the amount of living plant tissue in the circuit. This technique does not provide information on the distribution of the root system in the soil, only on the total amount of tissue. Work by Dalton (1987) is expanding the use of this technique to estimate root surface area.

Laboratory-oriented Non-Destructive Techniques

Although some of the field-oriented techniques can be applied to root systems grown in a controlled environment, most of the procedures described in this section are unique to laboratory studies. The tracer and capacitance techniques described above can be used in controlled situations, but the minirhizotron technique is not suitable for pot studies. Several variations of the rhizotron have been used in controlled environments. Generally these techniques are used for specific-purpose studies, require an increased equipment input, and are among the most recently developed procedures that we will describe.

Hydroponics and thin-film culture. These tools eliminate the soil matrix, which allows direct, nondestructive access to the roots. Measurements of total root production as well as detailed morphological studies are well suited to this technique. Hoagland and Arnon (1938) described the original development of this technique.

Plants are grown in tubes, chambers, or lined channels containing aerated nutrient solution. The nutrient balance of the solution imposes the desired conditions on the root system. The chemical environment surrounding the roots can be manipulated readily in these systems. Figure 4.7 shows a hydroponics system. Plants are held in position with a styrofoam block, suspending the root system in an aerated nutrient solution contained in plastic cylinders.

Aeroponics involves a system similar to hydroponics, but the solution injected into the root chamber is a mist rather than a flowing solution. Zobel, Del Tredici, and Torrey (1976) described an aeroponics system using a motor and spinner to create a fine mist. Thin-film culture involves a flowing layer of water a few millimeters thick that bathes the roots. The root system therefore grows horizontally on a plastic film, rather than suspended within a chamber.

A limitation of these techniques results from the characteristic that is its primary advantage. Roots are never exposed to the restraining force of a soil matrix. Therefore, a root system produced in this environment may not be identical to a soil-grown root system. Care must be taken in interpreting the results from experiments using hydroponics or thin-film culture.

Slant tubes or boxes. These techniques are variations of the rhizotron. They involve observing roots at the interface between soil and a clear window. Slant tubes differ from slant boxes only in the details of the container,

Figure 4.7. Hydroponic system in which plants are held in styrofoam blocks and roots are suspended in an aerated nutrient solution contained in plastic cylinders.

not in the principles of the technique. Slant tubes are clear cylinders filled with soil and mounted at an angle from vertical. Slant boxes are containers in which at least one wall is clear and inclined from vertical. The root observation surface is inclined to encourage root growth at the interface. A slant box system is shown in Figure 4.8.

Slant tubes have been used in screening programs to provide preliminary information about root development (Jordan 1983). The slant tube system is well suited for determining the rate of root elongation. Root extension can be measured at very small time intervals and the rate of elongation calculated. If the tube has sufficient length, the maximum rooting extension of a group of plants can be determined. These measurements of extension and the rate of elongation are relative rather than absolute values. Some rooting parameters will be affected by the small diameter of the tube, the properties of the inter-face, and the physical and chemical properties of the grow-ing medium, all of which can be substantially different from naturally occurring field conditions.

Slant boxes are well suited to the study of root responses to changes in the physical or chemical environ-ment. The root's response to intersecting a compacted lay-er or an acid subsoil can readily be seen on the wall of a slant box. Morphological studies can also be made using slant boxes. Stone and Taylor (1983) used a slant box ap-proach to study the effect of temperature on extension and orientation of soybean root systems.

Tracers. The same group of techniques described in the section on field-oriented techniques can be applied in a controlled environment with only slight modifications. The controlled environment may be better suited for using tracers than the field environment. Many tracers are either expensive or pose potential pollution risks and are not reasonable field tools. These same tracers, however, can be powerful laboratory tools.

Computer-aided Tomography (CAT). New and exciting tools for the study of root systems are emerging from technology developed for the medical industry, the CAT scanners. Several variations of three-dimensional imaging are being investigated. At least two different radiation sources, x-rays and gamma rays, are being used to measure the density of a matrix and thereby to infer the distribu-tion of water in that matrix. Because roots have a high concentration of water, they can be seen readily in the generated image. The distribution of water around individ-

Figure 4.8. Slant box system used in greenhouse studies. A removable glass panel, inclined 15S, is covered with an insulating light shield on one side of the box. The insulating light shield covers the glass panel except when root observations are being made.

ual roots can also be used in developing and validating refined theories of water flow (Hainsworth and Aylmore 1986).

At present, the primary limitations to the use of this
technology are expense, resolution, and availability of
equipment. The cost of the equipment can range into mil-
lions of dollars depending upon the specific requirements.
As the resolution increases, the equipment cost increases.
Resolution of these devices is limited by their design and
the energy source used in the measurements. Sample size,
soil water content, and the presence of certain elements
also limit the resolution. Increased sample size results
in lower resolution. At present, the resolution is a few
millimeters for a full root system. For smaller samples,
the resolution can be near 0.006 kg kg-1 water content over
distances of about 1.5 mm (Hainsworth and Aylmore 1986).
Energy limitations are related to the ability of the radiation to penetrate the soil matrix. As the thickness of the
matrix increases, the degree of attenuation increases and
significant scattering can occur, which will result in
lower resolution.

Nuclear magnetic resonance. Nuclear magnetic resonance (NMR) techniques were used by Sillerud and Heyser
(1984) to study the kinetics of sodium uptake and efflux.
They used dysprosium (III) tripolyphosphate as a paramag-
netic shift reagent. NMR techniques are also used to ac-
quire maps of water distribution in soil as affected by the
presence of plant root systems (Bottomly et al. 1986).

Neutron radiography. Neutron radiography has been
used to study the distribution and growth of roots in small
boxes of soil (Willatt et al. 1978). In this technique, an
aluminum box is filled with soil in which a seed is planted. At a desired time, the box is placed in a neutron beam
until an indium collector plate is exposed to about 5×10^9
neutrons cm^{-2}. The indium collector plate is transferred to
a high-contrast film, which is later developed. This technique was also used to study root diameter changes that
result from transpirational losses of water (Taylor and
Willatt 1983).

Fluorescence. A novel approach to the study of root
systems was used by Dyer and Brown (1983). The roots of
some lines of soybeans exhibit a natural fluorescence near
the apex of the root. Other lines do not exhibit this phenomenon. The rate of new root production is being studied
in growth membranes using this trait. If this trait is
found in other species it will allow expanded use of the
technique.

ROOT SYSTEM PARAMETERS

The various techniques for measuring the root system do not all measure the same parameter. Some techniques, such as minirhizotrons, rhizotrons, and trench profiles, involve counting roots at an interface. Other techniques, such as soil coring, provide a direct estimation of root length density, while tracer techniques yield an estimate of root activity. There is an apparent dichotomy between a physical parameter of size and root system activity. Measurements of root system size, however, are intended to act as surrogates for root activity or as an indicator of potential root activity.

Root length or root length density (length per unit soil volume) are the most commonly measured root parameters. This factor is the one most directly related to root function. It is the increase in root length that allows the root system to explore new soil volume.

Root surface area is more difficult to measure and interpret than root length but may be more directly related to root activity. Limited information is available on the effect on the activity of the root system of changing the root surface area, either the external surface area or that of the endodermis. Root surface area and root length, however, are often correlated.

Root biomass or dry weight is an appropriate measure of the photosynthetic input to the root system. Root biomass can be expressed on a whole plant basis or as a concentration of weight per unit soil volume. When carbon flow within the plant is of primary interest, biomass is the appropriate parameter.

CONCLUSIONS

In the past, many techniques have been developed for studying root systems, and new approaches are appearing regularly. The choice of the appropriate procedure and parameter in root research is critical to the outcome of the study. There are no clear guidelines, however, that can be applied in choosing the approach. It is clear that an effort should be made early in a project to define the type of root information needed and to identify the most appropriate procedure for collecting that information.

REFERENCES

Adeoye, K.B. and S.L. Rawlins. 1981. A split-root technique for measuring root water potential. Plant Physiology 68: 44-47.

Anghinoni, I. and S.A. Barber. 1980. Phosphorus influx and growth characteristics of corn roots as influenced by phosphorus supply. Agronomy Journal 72: 685-688.

Baker, J. M., and C. H. M. van Bavel. 1986. Resistance of plant roots to water loss. Agronomy Journal 78:641-644.

Bates, G. H. 1937. A device for the observation of root growth in the soil. Nature (London) 139:966-967.

Böhm, W. 1979. Methods of studying root systems. Berlin: Springer-Verlag.

Böhm, W., H. Maduakor, and H. M. Taylor. 1977. Comparison of five methods for characterizing soybean rooting density and development. Agronomy Journal 9:415-419.

Bottomly, P. A., H. H. Rogers, and T. H. Foster. 1986. NMR imaging shows water distribution and transport in plant root systems in situ. Proceedings of the National Academy of Science USA 83:87-89. Botany.

Chloupek, O. 1972. The relationship between electrical capacitance and some other parameters of plant roots. Biology Plantarium 14:227-230.

Dalton, F. W. 1987. In situ measurement of plant root area and mass. In Proceedings of International Conference. Measurement of soil and plant water status. 2:213-216. Logan: Utah State University.

Drew, M. C., and L. R. Saker. 1980. Assessment of a rapid method, using soil cores, for estimating the amount and distribution of crop roots in the field. Plant and Soil 55:297-305.

Dyer, D., and D. A. Brown. 1980. In situ root observation using fiber optic/video and fluorescence. p. 80. Agronomy Abstracts American Society of Agronomy.

Dyer, D., and D. A. Brown. 1983. Relationship between fluorescent intensity and ion uptake by soybeans. Plant and Soil 72:127-134.

Emanuelsson, J. 1984. Root growth and calcium uptake in relation to calcium concentration. Plant and Soil 78:325-334.

Foale, M. A. and D. R. Upchurch. 1982. Soil coring method for sites with restricted access. Agronomy Journal 74:761-763.

Fordham, R. 1972. Observations on the growth of roots and shoots of tea (Camellia sinensis L.) in Southern Malawi. Journal of Horticultural Science 47:221-229.

Freeman, B. M., and R. E. Smart. 1976. A root observation laboratory for studies with grapevines. American Journal of Enology and Viticulture 27:36-39.

Glover, J. 1967. The simultaneous growth of sugarcane roots and tops in relation to soil and climate. Proceedings of the South African Sugar Technology Association 41:143-159.

Goff, E. S. 1897. A study of roots of certain perennial plants. Wisconsin Agricultural Experiment Station Fourteenth Annual Report, 286-289.

Gregory, P. J. 1979. A periscope method for observing root growth and distribution in the field. Journal of Experimental Botany 30:205-214.

Hainsworth, J. M., and L. A. G. Aylmore. 1986. Water extraction by single plant roots. Soil Science Society of America Journal 30:841-848.

Hales, S. [1727] 1961. Vegetable statics. London. Reprint. London: MacDonald.

Hilton, R. J., D. S. Bhar, and G. F. Mason. 1969. A rhizotron for in situ root growth studies. Canadian Journal of Plant Science 49:101-104.

Hoagland, D. R. and D. I. Arnon. 1938. The water-culture method for growing plants without soil. University of California Agricultural Experiment Station Circular 347.

Huck, M. G., and H. M. Taylor. 1982. The rhizotron as a tool for root research. Advances in Agronomy 35:1-35.

Huxley, P. A., and A. Turk. 1967. A new root observation laboratory for coffee. Kenya Coffee. April 1967. pp. 156-159.

Jordan, W. R. 1983. Water relations of cotton. In Crop-Water Relations, ed. I.D. Teare and M. M. Peet. 213-254. New York: John Wiley & Sons.

Kaspar, T. C., D. G. Woolley, and H. M. Taylor. 1981. Temperature effect on the inclination of lateral roots of soybeans. Agronomy Journal 73:383-385.

Loneragan, J. F. 1979. The interface in relation to root function and growth. In The soil-root interface, ed. J. L. Harley and R. S. Russell. 351-367. London: Academic Press.

Lonkerd, W.E. and J.T. Ritchie. 1979. Split root observation system for root dynamics studies. Agronomy Journal 71:519-522.

Malik, R. S., J. S. Dhankar, and N. C. Turner. 1979. Influence of soil water deficits on root growth of cotton seedlings. Plant and Soil 53:109-115.

Mason, W. K., H. R. Rowse, A. T. P. Bennie, T. C. Kaspar,and H. M. Taylor. 1982. Responses of soybeans to two row spacings and two soil water levels. II. Water use, root growth and plant water status. Field Crops Research 5:15-29.

Mosher, P. N., and M. H. Miller. 1972. Influence of soil temperature on the geotropic response of corn roots (Zea mays L.). Agronomy Journal 64:459-462.

Nelson, W. W., and Allmaras, R. R. 1969. An improved mono- lith method for excavating and describing roots. Agronomy Journal 61:751-754.

Pearson, R. W. 1974. Significance of rooting pattern to crop production and some problems of root research. In The plant root and its environment, ed. E. W. Carson. 247-270. Charlottesville: University of Virginia Press.

Racz, G. J., D. A. Rennie, and W. L. Hutcheon. 1964. The P32 injection method for studying the root system of wheat. Canadian Journal of Soil Science 44:100-108.

Robertson, B. M., A. E. Hall, and K. W. Foster. 1985. A field technique for screening or genotypic differences in root growth. Crop Science 25:1084-1090.

Rogers, W. S. 1969. The East Malling root-observation laboratories. In Root growth, ed. W. J. Whittington. 361-376. London: Butterworth.

Rush, C. M., D. R. Upchurch, and T. J. Gerik. 1984. In situ observation of Phymatotrichum omnivorum with a borescope minirhizotron system. Phytopathology 74:104-105.

Sanders, J. L., and D. A. Brown. 1978. A new fiber optic technique for measuring root growth of soybeans under field conditions. Agronomy Journal 70:1073-1076.

Schuurman, J. J., and Goedewaagen, M. A. J. 1965. Methods for the examination of root systems and roots. Waggeningen: Pudoc.

Sillerud, L. D., and J. W. Heyser. 1984. Use of 23Na- nuclear magnetic resonance to follow sodium uptake and efflux in NaCl-adapted and non-adapted millet (Panicum miliacerum) suspensions. Plant Physiology 75:269-272.

Smucker, A. J. M., S. L. McBurney, and A. K. Srivastava. 1982. Quantitative separation of roots from compacted soil by the hydropneumatic alliteration system. Agronomy Journal 74:500-503.

Soileau, J. M., D. A. Mays, F. E. Khasawneh, and V. J. Kilmer. 1974. The rhizotron-lysimeter research facility at TVA, Muscle Shoals, Alabama. *Agronomy Journal* 66:828-832.

Stone, J. A., T. C. Kaspar, and H. M. Taylor. 1983. Predicting rooting depth as a function of soil temperature. *Agronomy Journal* 75:1050-1054.

Stone, J. A., and H. M. Taylor. 1983. Temperature and the development of the taproot and lateral roots of four indeterminate soybean cultivars. *Agronomy Journal* 75:613-618.

Stone, L. R., I. D. Teare, C. D. Nickell, W. C. Mayaki. 1976. Soybean root development and soil water depletion. *Agronomy Journal* 68:677-680.

Taylor, H. M. 1969. *The rhizotron at Auburn, Alabama - A plant root observation laboratory*. Auburn University Agricultural Experiment Station Circular 197.

Taylor, H. M., and S. T. Willatt. 1983. Shrinkage of soybean roots. *Agronomy Journal* 75:818-820.

Taylor, H. M., and S. T. Willatt. 1981. Utilization of rhizotrons in root research. In *The soil/root system in relations to Brazilian agriculture*, ed R.S. Russell, K. Igue, and Y. R. Mehta. 319-337. Londrina, Brazil: IAPAR.

Taylor, H. M., ed. 1987. *Minirhizotron observation tubes: Methods and applications for measuring rhizosphere dynamics*. ASA Special Publication no. 50. Madison,WI: American Society of Agronomy.

Upchurch, D. R. 1985. Relationship between observations in minirhizotrons and true root length density. Ph.D. dissertation. Lubbock: Texas Tech University Press.

Upchurch, D. R., B. L. McMichael, and H. M. Taylor. 1988. Use of minirhizotrons to characterize root system orientation. *Soil Science Society of America Journal* 52:319-323.

Upchurch, D. R., and J. T. Ritchie. 1983. Root observations using a video recording system in minirhizotrons. *Agronomy Journal* 75:1009-1015.

Upchurch, D. R., and J. T. Ritchie. 1984. Battery-operated color video camera for root observations in minirhizotrons. *Agronomy Journal* 76:1015-1017.

Waddington, J. 1971. Observations of plant roots *in situ*. *Canadian Journal of Botany* 49:1850-1852.

Weaver, J. E. 1919. *The ecological relations of roots*. Carnegie Institute of Washington Publication no. 286. Washington, DC.

Weaver, J. E. 1926. <u>Root development of field crops</u>. New York: McGraw-Hill Book Co.

Willatt, S. T., R. G. Struss, and H. M. Taylor. 1978. <u>In situ</u> root studies using neutron radiography. <u>Agronomy Journal</u> 70:581-586.

Zobel, R. W., P. Del Tredici, and J. G. Torrey. 1976. Method for growing plants aeroponically. <u>Plant Physiology</u> 57:344-346.

P. G. Hunt

5 Microbial Responses in the Rhizosphere of Agricultural Plants

The interactive relationship between plant roots and microorganisms was recognized and expressed in 1904 by Lorenz Hiltner, soil bacteriologist and professor of agronomy at the Technical College of Munich. In early studies, plant pathogenic and symbiotic processes that produced morphological changes visible to the unaided eye were the most intensively investigated relationships. Today, study of microbial ecology in plant rhizospheres is extremely broad and complex. Plant and microbial relationships are, of course, affected by the total biophysical system of both the plant and the soil. The system might be viewed as progressive development of interactions among soil structure, growing roots, rhizosphere microorganisms, soil and canopy environments, and the physiology of the entire plant. Consequently, the holistic understanding of rhizosphere microbial ecology has become vitally important to soil management and plant growth.

Other chapters in this volume focus on specific aspects such as soil structure, mycorrhizae, plant pathogenicities, nitrogen cycling, soil fauna, and shoot/root development. The emphasis of this chapter will be microbial involvement in soil structure and nitrogen fixation in the rhizospheres of plants grown in agricultural systems. Only a few of many studies will be used to illustrate particular points; thus, some excellent investigations may not be cited.

RHIZOSPHERE ESTABLISHMENT AND SOIL STRUCTURE

Root growth is fast in young seedlings. The pattern, rate, and extent of root development is, of course, dependent upon the genetic makeup of the plant and environmental conditions. For instance, the rooting forces necessary to penetrate soil layers are directly related to the soil strength characteristics, which are dramatically affected by soil structure and water content. Rooting in many soils may occur only when the soil is wet enough to promote low strength characteristics but dry enough to have adequate O_2 contents (Campbell, Reicosky, and Doty 1974; Campbell and Phene 1977). Thus, microbial promotion of soil structure prior to, as well as in conjunction with, root growth is very important.

The production of extracellular polysaccharides is predominantly a bacterial process, and it promotes what can be referred to as the "cementing" of particles. Chaney and Swift (1986) reported that soil aggregates were reformed by polysaccharides produced from glucose amendment but not by wetting/drying and freezing/thawing cycles. These aggregates declined over a twelve-week period as polysaccharides were decomposed. The stability of these aggregates was related to the original organic matter levels in the soils. Chapman and Lynch (1984) found that the polysaccharides of straw were composed mainly of galactose, glucose, and mannose. The ability of straw to increase aggregation was also shown to be inversely related to straw nitrogen contents between 0.25 and 1.09 percent by Elliott and Lynch (1984).

The use of bacteria and fungi in a coculture was described by Chapman and Lynch (1985). They found increased aggregate stability of a Humble silt loam with the coinoculation of Trichoderma harzianum and Enterobacter cloacae. They found that the cellulolytic fungi could support the growth of bacteria and supply sufficient carbon for polysaccharide synthesis. Anaerobic and facultative anaerobic nitrogen-fixing bacteria were thought to be sustained by cellulolytic enzymes of aerobic fungi (Harper and Lynch 1984). Fungi play a major role in the stability of soil aggregates as their mycelia grow around soil particles and ridge to others. This bridging is of increased importance once the root grows near the soil particles. Fungi also play a role in the transport of chemicals among microsites (Newman 1985).

Once the root occupies an area of the soil, it affects and is affected by the microorganisms of that region. Thus, the root has an immediate effect on the microbial ecology of the soil (Figure 5.1). Roots excrete large amounts of photosynthate (Lynch et al. 1981), and these organic compounds serve as substrates for microbial populations. The plant roots and shoots are, in turn, affected by the microbes. For example, Barber and Lynch (1977) found that the biomass of bacteria produced in the rhizosphere of barley seedlings was greater than could have been produced from the carbohydrates released in sterilized soil. These data supported the view that microbes do not simply grow on plant roots; they stimulate plants to release more photosynthate. Barber and Martin (1976) estimated that the exudate losses of wheat (_Triticum aestivum_) and barley (_Hordeum vulgare_) seedlings were equivalent to 7 to 13

I. Initial Root Growth

II. Secretion of Organic Matter and Proliferation of Bacteria and Fungi

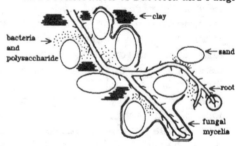

III. Establishment of Associative and Symbiotic Bacteria

IV. Shoot–Root–Bacteria Dynamics

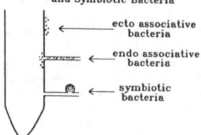
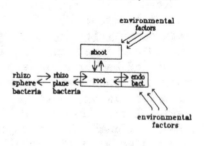

Figure 5.1. A schematic of root development, rhizosphere establishment, and shoot-root-microbial equilibrium.

percent and 18 to 25 percent of their dry matter under
sterile and nonsterile conditions, respectively. Microbially produced growth regulators may also stimulate increased total growth of the plant. Beck and Gilmour (1983)
found that wheat released 3.7 and 3.0 percent of the [14]C-
labeled photosynthate as soluble exudates when inoculated
and noninoculated with nitrogenase-positive bacteria, respectively. Whipps and Lynch (1983) concluded that competition for available root exudate on the root of barley would
be high and that it would preclude luxuriant bacterial
growth in the rhizosphere.

The degree of the interrelation between the root and
microbe varies. Those microorganisms that live within or
in close association with the root are most directly
affected by the plant. This region can be called the
endorhizosphere. Many microbes also occupy the rhizoplane,
which is the actual root surface that contacts the soil,
water, and air. The third zone is the exorhizosphere,
which is separated from the root by a film of water. A
good view of these environments was expressed by Whipps
and Lynch (1983). They suggested that regions can best be
viewed as a continuum that is by nature dynamic with plant
growth.

In addition to microbial effects, physical and chemical factors such as soil texture, structure, color, organic
content, water status, pH, and salinity play major roles in
the expression of rooting potential. These factors along
with variations in plant canopy condition make the rhizosphere quite variable from plant to plant as well as from
field to field.

ASSOCIATIVE NITROGEN FIXATION

Associative nitrogen-fixing microorganisms have been a
source of interest for many years. Earlier work with Azotobacter inoculation has been controversial (Brown 1972).
Several rhizoplane genera such as Bacillus, Enterobacter,
Pseudomonas, and Beijerinckia are capable of nitrogen fixation in the rhizosphere of grain crops such as corn (Zea
mays), wheat, sorghum (Sorghum bicolor), and rice (Oryza
sativa) as well as in the rhizosphere of tropical forages
(Neyra and Dobereiner 1977; Hubbell and Gaskins 1984).
Dobereiner (1961) was one of the first to galvanize recent
thought about the use of associative nitrogen fixation when
she showed that Beijerinckia was stimulated by the sugar-

cane (<u>Saccharum officinarum</u>) rhizosphere. Subsequently, Dobereiner and Day (1976) reported nitrogen fixation in the rhizosphere of <u>Digitaria decumbes</u> by the bacterium now known as <u>Azospirillum</u>.

A complete discussion of associative nitrogen fixation is far beyond the scope of this chapter. The genus <u>Azospirillum</u> has received a great deal of world-wide attention in recent years. Its two well-studied species, <u>A. brasilense</u> and <u>A. lipoferum</u>, are good generic examples of bacteria that grow in the rhizosphere, fix nitrogen, and produce plant growth regulators.

Occurrence of Azospirillum

In field and greenhouse work, Baldani and Dobereiner (1980) found that <u>A. lipoferum</u> was much more commonly isolated from externally sterilized roots of maize (<u>Zea mays</u>) than from wheat, but the reverse was true for <u>A. brasilense</u>. In Hawaii, Kosslak and Bohlool (1983) found <u>A. brasilense</u> and <u>A. lipoferum</u> to be present in a number of plants in about equal percentages. In their review of associative nitrogen fixation by <u>Azospirillum</u>, Hubbell and Gaskins (1984) suggested that <u>Azospirillum</u> was generally found in associations with plants where investigators looked for them and that the frequent discovery of <u>Azospirillum</u> on the roots of tropical grasses might be more of a quantitative than a qualitative nature.

Plant Responses to Associative Bacteria

O'Hara, Davey, and Lucas (1981) reported that seven strains of <u>A. brasilense</u> increased the dry matter and nitrogen content of maize shoots when they were grown under temperate conditions in sand-filled vermiculite pots. The positive response of plant growth to <u>A. brasilense</u> inoculation in India has been reported by Rai (1985a,b,c). The positive response in seed yields were particularly evident when soil and fertilizer nitrogen were low. Strains of <u>A. brasilense</u> that were adapted to the prevailing soil environments were most effective in increasing corn, cheena (<u>Panicum millaceum</u>), and millet (<u>Eleusine coracana</u>) growth and yield. Also in India, Meshram and Shende (1982) reported increased grain yield as well as increased nitrogen content in maize grain and stover after the inoculation of maize with <u>A. chroococcum</u>. Rai and Gaur (1982) reported that maize yields were 2.97 t/ha with nitrogen treatment of 80 kg/ha and 4.15 t/ha when A. lipoferum inoculant was used along with the same fertilizer nitrogen. The particular strain of A. lipoferum had a high nitrogen-fixing capacity and no denitrifying tendency.

Smith et al. (1976) reported that several tropical grasses and cereals of North America experienced enhanced growth when inoculated with Spirillum lipoferum. The growth and total nitrogen content of wheat, sorghum, sorghum X Sudan grass (Sorghum bicolor Sudanense), and Proso millet Panicum millaceum) were increased by inoculation with A. brasilense in Israel (Kapulnik et al. 1981a). Yield of summer cereal crops and Setaria italica were also increased by inoculation with Azospirillum (Kapulnik et al. 1981b).

Yields of winter and spring wheat were increased by aerial application of A. brasilense in field studies during 1979 and 1980 (Reynders and Vlassak 1982). They believed that the bacteria caused increased tillering and nutrient uptake. Sarig et al. (1984) reported increases in both grain and forage yield for sorghum cultivars grown under dryland conditions in Israel. Smith et al. (1976) reported increases in the herbage produced by Pearl millet (Pennisetum americanum) and Guinea grass (Panicum maximum) when they were inoculated with A. brasilense, but a majority of forty tropical grasses tested did not show increases in growth.

Nitrogen Fixation

The amount of nitrogen obtained from associative fixation by Azospirillum spp. has been estimated at different levels, and some of the very high estimates are no doubt due to an overestimation bias of the acetylene reduction method (Gaskins and Carter 1975; van Berkum and Bohlool 1980; von Bulow and Dobereiner 1975). Apparent overestimations have occurred in other associations. Brown (1976), for example, also reported that the main advantage of Azotobacter paspali in the rhizosphere of Paspalum notatum was the production of growth regulators rather than nitrogen fixation. This is in contrast to estimates of as much as 90 kg/ha nitrogen fixation by A. paspali on P. notatum by Dobereiner, Day, and Dart (1972).

Production of Growth Regulator Substances

The low nitrogen fixation rates of associative nitrogen-fixing bacteria support the view that much, if not all, of the plant growth stimulation derived from associative nitrogen-fixing bacteria comes from plant growth hormones rather than nitrogen fixation (Tien, Gaskins, and Hubbell 1979; Tien et al. 1981; Lin, Okon, and Hardy 1983). In either case, it appears that associative nitrogen-fixing bacteria can often increase the yield of important forages and grains, especially when soil fertility is low and the strains are adapted to restrictive environmental conditions such as pH, salinity, or temperature.

SYMBIOTIC NITROGEN FIXATION

Biological nitrogen fixation is certainly among the most important rhizosphere processes. Symbiotic nitrogen fixation was a cornerstone of soil fertility until the development of chemical conversion of atmospheric nitrogen to ammonia; chemical conversion gave the world a new source of nitrogen fertilizer which greatly increased agricultural productivity. Nitrogen fertilizers are not, however, panaceas. Conversion of atmospheric N^2 to ammonium is an energy-intensive process, and fertilizers so produced must be transported to locations of agricultural use. This transportation is expensive and energy consumptive at best, and not available in many areas of the world. Additionally, overuse of nitrogen in areas such as the midwestern section of the United States has led to groundwater contamination. Thus, there is continued interest in legume-rhizobial symbiosis.

For an effective legume-rhizobial symbiosis, the rhizobia must be present, infection/nodule formation must occur, and the rhizobia must be an efficient nitrogen fixer within the plant. Rhizosphere conditions impact all of these critical aspects of dinitrogen fixation.

Rhizobial Presence and Survival

Rhizobia that are capable of infecting and forming nodules in a particular legume must be present or introduced into a soil if effective symbiosis is to be established. Cropping and farm management practices will, of course, have dramatic impact on the number of rhizobia present. For example, Weaver, Frederick, and Dumenil (1972) determined that the number of Bradyrhizobium japonicum cells in fifty two Iowa fields was positively correlated to the presence of soybean (Glycine max) in the crop rotation during the previous thirteen years. Hiltbold, Patterson, and Reed (1985) found that the rhizobial numbers of B. japonicum were high (10^6 cell g^{-1}) during the winter following soybean production. If lime, phosphorus, and potassium were adequate, rhizobial numbers stabilized around 10^4 or 10^5 cells g^{-1} of soil. Soil acidity reduced rhizobial populations. When pH levels were less than 4.6, rhizobial numbers were generally less than 10^2 g^{-1} of soil. Rupela et al. (1987) found that research station soils in India contained from 10^3 to 10^5 rhizobia cells g^{-1} of soil, while those of farmers contained from 10 to 10^3 rhizobia g^{-1} of soil. Populations of rhizobia were highest during and shortly after the growth of a legume crop. Populations then decreased to a lower level, but levels were generally quite sufficient for inoculation of subsequent crops unless extreme environmental conditions were present.

Rhizobial populations can be greatly affected by factors such as water content (Pena-Cabriales and Alexander 1979), temperature (Munevar and Wollum 1981; Osa-Afiana and Alexander 1982; Kvien and Ham 1985), pH (Keyser and Munns 1979b), salinity (Rai et al. 1985; Rai 1987), and nutrient status of the soil (Keyser and Munns 1979a,b). As a result, some areas planted to legumes may not be high in numbers of rhizobia or percentage of effective nodulators. For example, cowpea (Vigna unguiculata) rhizobial populations of Guyana soils were low enough in both numbers and effectiveness for potential benefit from inoculation with effective strains (Trotman and Weaver 1986). Rao et al. (1985) reported that rhizobia used to inoculate the American soybean cultivar "Bossier" in Nigerian soils were found in soybean grown after two years of fallow, but the yield of the soybean was increased by annual inoculation. Survival of the rhizobia may even be inhibited by the seed as in the case of R. trifolii and arrowleaf clover (Trifolium vesiculosum) (Materon and Weaver 1984). Fuhrmann, Davey, and Wollum (1986) found differences in the effects of desiccation on R. leguminosarum bv. trifolii on a Altavista loamy sand and a Cecil sandy clay loam. Population levels were generally lowered by incubation at the -70 MPa level, but one isolate had excellent survival under all soil moisture treatments, including the -500 MPa moisture level. Wollum and Cassel (1984) used geostatistical techniques to assess the spatial variability of rhizobial populations. Large population variations occur not only from field to field and soil to soil but over small distances in the same field and soil.

Infection and Nodule Formation

Rhizobia have been classified by their ability to inoculate certain plants, but this has not been entirely satisfactory. Certain rhizobia are capable of infecting and nodulating plants of several species, and certain plants are nonselective in their nodulation. In addition, rhizobia have been classified by their rate of growth on synthetic media, i.e., fast- or slow-growing (Jordan 1982; Sadowsky, Keyser, and Bohlool 1983). It is also important to understand that infection involves both the plant and the microorganism. Even after recognition and infection by the rhizobia, plants can regulate subsequent nodulation by the same or other rhizobia (Kosslak and Bohlool 1983), and this regulation can be affected by environmental conditions as subtle as the spectral composition of canopy light (Kasperbauer, Hunt, and Sojka 1984; Hunt, Kasperbauer, and Matheny 1987).

One of the most difficult problems associated with improvements in the legume/rhizobia symbiosis in agricultural systems is the low competitiveness of introduced strains of rhizobia relative to indigenous strains. This is true even if the introduced strain will readily infect and form nodules on a plant when the strain is present in a single culture. Differences in infection potential can be seen from the fact that serogroup 123 dominated soybean nodules with 60 to 100 percent occupancy, even though it did not dominate the rhizobial population of the rhizosphere; the numbers of B. japonicum serogroups 110, 123, and 138 all increased in the rhizosphere of soybean to about 10^6 cells g^{-1} soil (Moawad, Ellis, and Schmidt 1984). Populations of serogroups 110, 123, and 138 in fallow soil were each about 10^5 cells g^{-1} soil. Moawad and Bohlool (1984) found that strain B214 was least competitive among six strains for nodulation of Leucaena leucocephala in an oxisol (less than 30 percent) but most competitive in a mollisol (70 percent). The correlation between occupation of nodules and rhizosphere populations was low. Competitive advantages of various rhizobia not only change with soils, but with tillage and cultivar on the same soil (Hunt, Matheny, and Wollum 1985). They concluded that tillage, cultivar, and inoculation interacted to influence B. japonicum strain occupancy (Table 5.1).

Weaver and Frederick (1974) found that introduced strains must be present in numbers 10^3 or 10^4 greater than the indigenous strains to significantly impact nodule occupancy of soybean. Even these high numbers often produced less than 10 percent nodule occupancy in field inoculation studies in the southeastern United States (Hunt et al. 1983). Dunigan et al. (1984) initially found that infection of soybean by B. japonicum strain USDA 110 was low (0 to 17 percent) even when massive inoculations (10^8 cells cm^{-1}) of B. japonicum were used for three consecutive years. During the subsequent four years, however, the percentage of nodules infected by the introduced strain increased to as high as 60 percent; the researchers interpreted this to indicate that a prolonged application of massive inoculation would result in the establishment of the introduced strain as a significant portion of the rhizobial population. It is possible that prolonged use of quality inoculum may become an important practice, but substantial progress must be made in the uniformity and quality of inoculum (Giddens, Dunigan, and Weaver 1982).

Viteri and Schmidt (1987) reported that indigenous soil rhizobia could respond to the addition of several sugars, particularly arabinose; B. japonicum populations increased from about 10^4 to about 10^6. They interpreted this as evidence that indigenous populations of rhizobia could respond to various substrates in the absence as well as in the presence of a host plant.

Trinick, Rhodes, and Galbraith (1983) found that the fast- and slow-growing rhizobia competed differently under temperatures of 25 and 30° C. They also reported a difference in competitiveness with variation in the day/night temperatures. This is of considerable importance because it represents the normal fluctuation of the daily temperature cycle and the differences that exist with different soil and water management systems, i.e. crop residue management, water management, and tillage. Rai and Gaur (1982) and Rai (1987) reported selection of lentil rhizobia that had improved temperature, pH, and salinity tolerance. Mahler and Wollum (1981) reported differences in the drought tolerance of B. japonicum strain in the symbiotic as well as in the free-living state.

Differences also are found among strains for phage tolerance. A rhizobiophage of B. japonicum USDA 117 was found to reduce the nodule number, nodule weight, and acetylene-reduction capacity of plants inoculated with USDA 110

TABLE 5.1. Nodular occupancy by eight Bradyrhizobium japonicum strains as affected by soybean cultivar, tillage, and inoculation with strain 110 of B. japonicum.

Tillage	Inocu- lation	Bradyrhizobium japonicum strains							
		24	31	46	76	94	110	122	125
		------------------------------%-----------------------------							
1980									
					Lee				
Conventional	+	3	7	4	3	3	7	8	2
	-	2	11	1	8	5a	4	8	0
Conservation	+	1	6	1	8	10b	5	3	6a
	-	1	13	5	7	0ab	0	6	1a
					Ransom				
Conventional	+	1	22	4	7	0	12	6	0
	-	0	16	2a	7	0	9	4	0
Conservation	+	3	21	4	6	0	7	5b	0
	-	0	28	9a	4	0	3	0b	0
					Coker				
Conventional	+	1	10	8	12	9	3	4	2
	-	0	19	6	8	14	4	4	0
Conservation	+	0	13	3	12	11	2	11	1
	-	0	16	9	7	10	2	4	1
CVd		77	35	39	31	44	56	48	59

1981

				Lee					
Conventional	+	9	8	6a	1b	9b	2	13	0
	−	13	6	5	4b	20b	0	8	0
Conservation	+	11	4	0ab	0c	20	4b	6	0
	−	20	2	4b	4c	10	0b	7	0
					Ransom				
Conventional	+	7	14b	6	1ab	14	2	3ab	0
	−	5	5b	7	5b	26	0	11b	0
Conservation	+	9	14	6	6a	11	0b	8a	0
	−	12	8	6	7	12	3b	9	0
					Coker				
Conventional	+	13a	7a	6b	3	12	1a	11	1
	−	15	6a	1b	2	18	0	6	0
Conservation	+	15	6a	7c	8b	26	6ab	10	0
	−	10b	15ab	1c	3b	16	2b	8	0
CVd		36	30	41	44	31	60	28	29

a,b,c   Means for the same year, serogroup, and cultivar followed by "a" are
different for tillage and those means followed by "b" or "c" are
different for inoculation by the LSD test at the 0.10 level when
analyzed after a square root of (mean + 0.5) transformation.

d   CV values were calculated from transformed data; therefore, means
within a column should be transformed by a square root of (mean + 0.5)
before comparing to CV values.

Source: Hunt, Matheny, and Wollum 1985.

(Hashem and Angle 1988).

Iron deficiency was found to inhibit the nodule development of ground nut (<u>Anachis hypogaea</u>) in calcareous soil of Thailand (O'Hara et al. 1988). However, iron deficiency did not limit growth or populations of ground nut Brady-rhizobia in the soil or rhizosphere. Whelan and Alexander (1986) found that <u>R. trifolii</u> did not nodulate subterranean clover (<u>Trifolium subterraneum</u>) in the presence of high levels of iron or aluminum (500 and 50 micromoles, respectively) nor below pH 4.8. Riley and Dilworth (1985) reported that the adverse effect of cobalt deficiency on nitrogen fixation in <u>Lupinus angustifolius</u> was due to the inability of the plant to supply cobalt to the rhizobia rather than an effect on their growth in soil.

Many postulates have been made about why introduced strains are less competitive than indigenous strains. Greater environmental tolerance, more homogenous distribution in the soil, and improved recognition abilities are possible reasons. Selection of plants that will exclude predominant rhizobial strains has been done (Cregan and Keyser 1986), but there are many questions that must be answered before such a process can be used agronomically. At least partial resolution of the problem could rest with improved planting environments and inoculants, i.e., favorable soil water status, neutral pH, low salinity, adequate nutrition, and moderate temperature, along with large numbers of viable cells in the inoculum.

Efficient Nitrogen Fixation

Young, Hughes, and Mytton (1986) found dramatic increases in dry matter production during the first year after inoculation of white clover (<u>Trifolium repens</u>) with <u>R. trifolii</u>; and such results are not uncommon. Yet, strains of rhizobia that carry out efficient nitrogen fixation in one plant may fail to do so in another. In some plants, infection and nodulation occur, but nitrogen fixation is nonexistent or ineffective (Mathis, Kuykendall, and Elkan 1986). Under these conditions, the rhizobia acts as a parasite. It is using photosynthate to grow, but it is providing no nitrogen for plant growth. Keyser et al. (1982) reported a very interesting ineffective infection of fast-growing rhizobia (generation times of two to four hours) with soybean. They isolated these rhizobia from wild progenitor soybean (<u>Glycine soja</u>) in China. Soybean (<u>Glycine max</u>) is normally only infected by slow-growing rhizobia (generation times of more than six hours). The nodules formed by the fast-growing rhizobia were only effective in the cultivar "Peking," a black-seeded, genetically unimproved line from China.

There also exists distinct differences in the nitrogen fixing capacities of legume species. Differences between a forage legume such as alfalfa (Medicago sativa) and the common bean (Phaseolus vulgaris) are generally perceived. There also exist more subtle differences among grain legumes; Piha and Munns (1987a,b) found that common beans obtain relatively less nitrogen from fixation than soybean or cowpea. The same was true for several Phaseolus spp., and Piha and Munns concluded that common bean species may be genetically predisposed to lower nitrogen fixation.

The contribution of legumes to the annual nitrogen balance of a soil depends upon several factors, such as the crop, soil type, rhizobial effectiveness, rainfall pattern, residue management, seed yield, and seed nitrogen content. If a crop such as alfalfa is plowed under rather than harvested, the addition of nitrogen to the soil can be several hundred kg/ha (Heichel 1987). If it is harvested several times, however, it may in fact be a net consumer of soil nitrogen. Soybean grown in the midwestern United States is generally a net annual consumer of nitrogen, while that grown in the southeastern United States is generally a net nitrogen producer (Welch et al. 1973; Hunt, Matheny, and Wollum 1985; Thurlow and Hiltbold 1985). This is predominantly the result of soil-nitrogen differences; naturally low levels of soil nitrogen in the southeastern United States allow the fixation and accumulation of large amounts of nitrogen in legumes.

Plant Growth and Seed Yield

If nitrogen is the limiting plant growth and seed yield factor, establishment of a more effective nitrogen-fixing symbiosis will definitely improve seed yield. This has been shown by the marked increase in dry matter and seed yield of nodulating relative to nonnodulating soybean grown in the nitrogen-limiting soils of the southeastern United States (Matheny and Hunt 1983; Thurlow and Hiltbold 1985; Hunt, Matheny, and Wollum 1985). The amount of nitrogen obtained from fixation, however, can be increased without an increase in seed yield (Morris and Weaver 1983; Williams and Phillips 1983), and yield can be increased from rhizobial inoculation without increased nitrogen fixation (Karlen and Hunt 1985; Hunt, Matheny, and Wollum 1985; Hunt et al. 1985). This is possible because the rhizobia are in an interactive state with the plants, other micro-organisms, and the environment as depicted in phase IV of Figure 5.1. Under these conditions, the rhizobia are affecting the reaction of plants to the environment by means other than nitrogen fixation, most likely via growth-regulating hormones. This may be more evident in rhizobia with the capacity to produce higher levels of hormones such as indole-3-acetic acid (IAA), but initial field studies with such rhizobia indicate that they do not greatly alter the growth of soybean under field conditions (Kaneshiro and Kwolek 1985; Hunt, Kaneshiro, and Matheny 1987). However, the addition of a low concentration of precursor to the rhizosphere of plants inoculated with growth regulator-producing microorganisms, may cause substantial growth alterations (Frankenberger and Poth 1987). It is even possible that rhizobia may stimulate the germination of nonlegumenous plants such as wheat or corn (Kavimandan 1986).

Interactions of B. japonicum strain with irrigation have been reported (Hunt, Wollum, and Matheny 1981, Hunt et al. 1983, and Hunt, Matheny, and Wollum 1988 (Table 5.2). The soybean seed yield response to irrigation varied with cultivar and strain. They also found differences in seed yield of soybean grown under drought conditions when the soybean was inoculated with cultures of the B. japonicum strain USDA 110 that had been maintained in different laboratories. The effects of canopy configuration (row width and compass orientation) of soybean were accentuated or diminished by the strain of B. japonicum (Hunt et al. 1985, 1990) (Figure 5.2).

Row orientation and strain of rhizobium were also important to the nitrogen fixation and seed yield of chickpea (Cicer arietinum) in India (R. Rai, PL-480 report and personal discussions 1988). The effects of row orientation and soil color on soybean nodulation were postulated to be related to the spectral composition of canopy light, which affected shoot:root ratios, the extent of nodulation, and the relative competitiveness of various strains (Kasperbauer and Hunt 1987, 1988; Kasperbauer, Hunt, and Sojka 1984; Hunt, Kasperbauer, and Matheny 1987, 1989) (Table 5.3). The autoregulation of nodulation expressed by one side of a split root system on the other could be increased or decreased by Red (R) or Far-red (FR) end-of-day light treatment. Since FR effect could be reversed by R, the autoregulation was partially controlled by the plants phytochrome system. Additionally, when soybean were inoculated

with <u>B. japonicum</u> strains USDA 110 or Brazil 587 and treated with (R) or (FR) light nodulation was significantly
altered if the inoculant was USDA 110 but not if the inoculant was Brazil 587 (Hunt et al. 1990). Thus, the light
quality environment of the shoot is able to effect the
relative competitiveness of rhizobia in the rhizosphere.
It is possible that some of the more important advances in
legume seed yield may come from a better understanding of
the whole plant response to the rhizobia rather than simple
improvements in nitrogen fixation efficiency.

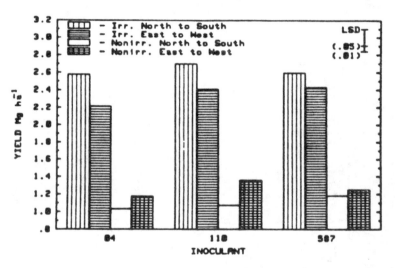

Figure 5.2. Coker 338 soybean yield response to row
orientation, inoculation, and irrigation in 1982.

SUMMARY

In summary, it has been the author's intent in this
chapter to show the close, continual, and dynamic
relationships that exist among plants, rhizosphere
microorganisms, and the environment, with emphasis on
rhizosphere establishment and nitrogen fixation processes.
These interrelations are somewhat cyclic and may be thought
of as starting when soil microorganisms begin to degrade
crop residues, produce polysaccharides, and build soil
structure. Soil structure is important to the rate and
extent of root development in subsequent crops, and the
interaction of rhizosphere bacteria and fungi are critical

TABLE 5.2. Two-year mean seed yield and seed yield rank for soybean inoculated with different B. japonicum strain and grown under irrigated (+) and nonirrigated (-) conditions in 1981 and 1983.

Cultivar	Irr.	Bradyrhizobium japonicum strains									
		1001	1004	1005	1010	1029	110	587	Nat	T102	TI'84
		------Mg ha-1------									
Cobb	-	1.08	1.50	.89	1.17	1.28	.98	.85	.93	.83	1.03
	+	1.68	1.40	1.74	1.97	1.96	1.50	1.72	1.75	2.01	1.80
Coker 448	-	1.19	1.47	1.89	1.97	1.75	1.60	1.46	1.17	1.12	1.28
	+	2.32	2.94	2.38	3.03	2.07	2.92	2.53	2.59	2.52	2.37

LSD(0.05)=0.59
LSD(0.01)=0.77

Cultivar	Irr.	------Rank------									
Cobb	-	101	143	78	110	111	92	78	85	77	97
	+	146	129	162	179	178	137	153	156	172	165
Coker 448	-	109	132	163	171	158	139	122	98	92	113
	+	186	258	204	270	184	253	235	225	212	192

LSD(0.05)=57
LSD(0.01)=75

Source: Hunt, Matheny, and Wollum 1988.

TABLE 5.3. Effects of red (R) and (far-red) FR light treatment of shoots and root inoculation time on nodule number for soybean grown in a split-root system.

Inoculation schedule[b]	Light treatment					
	R		FR		FR,R	
	Side		Side		Side	
	A	B	A	B	A	B
	No./plant*					
A_0B_0	17 ± 4^a	17 ± 3^a	8 ± 2^a	11 ± 1^a	11 ± 4^a	12 ± 6^a
$A_{6-9}B_{6-9}$	27 ± 6^a	22 ± 4^a	21 ± 5^a	19 ± 4^a	24 ± 5^a	18 ± 5^a
A_0B_{2-4}	32 ± 5^a	14 ± 3^a	14 ± 3^a	5 ± 2^a	18 ± 4^a	11 ± 3^a
A_0B_{4-6}	38 ± 5^a	7 ± 2^a	16 ± 2^a	2 ± 1	29 ± 8^a	4 ± 3^a
A_0B_{6-9}	38 ± 7^a	2 ± 2	21 ± 4^a	0 ± 0	32 ± 8^a	0 ± 0
A_0B_9	39 ± 8^a	0 ± 0	22 ± 5^a	0 ± 0	26 ± 5^a	0 ± 0
Noninoc	0 ± 0	0 ± 0	0 ± 0	0 ± 0	0 ± 0	0 ± 0

*Values are means ± SE.

[a] Numbers within the same column are significantly different from the noninoculated control at P compared by a single degree of freedom contrast. Numbers were transformed by Box-Cox transformation = 0.11 for homogeneity of variance before analysis.

[b] Letters indicate side of the split root and numbers indicate inoculation times in days after time zero.

Source: Hunt, Kasperbauer, and Matheny 1987.

aspects of enhanced soil structure during the crop growth periods. Once the root grows into the soil microsites, rhizosphere bacteria stimulate production of root exudate, which in turn is used by the microbes to produce more growth-stimulating materials and polysaccharides that can cement and stabilize soil particles. Fungi continue the process by bridging soil aggregates and roots. Associative and symbiotic bacteria become established and promote plant growth by growth regulator production as well as nitrogen fixation. At this point, interactions of the plant, rhizosphere microorganism, and environment of both the shoot and root are dynamic; changes in one part of the system affect all parts of the system. For instance, plant responses to changes in factors such as water status and spectral composition of canopy light are interactive with the presence and competitiveness of rhizosphere rhizobia.

This dynamic state of continual interaction and adjustment is representative of how microorganisms function and respond in the rhizosphere of agricultural crops. Much progress has been made in the nearly 90 years since Professor Hiltner first discussed these interactive relationships. Yet, we have only begun to understand and apply rhizosphere concepts to modern agriculture. It is indeed possible that discovery and application of new information in the microbial aspects of the rhizosphere will be vital building blocks for profitable and environmentally sustainable agriculture enterprises of the future.

REFERENCES

Baldani, V. L. D., and J. Dobereiner. 1980. Host-plant specificity in the infection of cereals with Azospirillum spp. Soil Biology and Biochemistry 12:433-439.

Barber, D. A., and J. M. Lynch. 1977. Microbial growth in the rhizosphere. Soil Biology and Biochemistry 9:305-308.

Barber, D. A., and J. K. Martin. 1976. The release of organic substances by cereal roots into soil. New Phytology 76:69-80.

Beck, S. M., and C. M. Gilmour. 1983. Role of wheat exudates in associative nitrogen fixation. Soil Biology and Biochemistry 15:33-38.

Brown, M. E. 1972. Plant growth substances produced by micro-organisms of soil and rhizosphere. Journal of Applied Bacteriology 35:443-451.

Brown, M. E. 1976. Role of Azotobacter paspali in association with Paspalum notatum. Journal of Applied Bacteriology 40:341-348.

Campbell, R. B., and C. J. Phene. 1977. Tillage, matric potential, oxygen, and millet yield relations in layered soils. Transactions of the American Society of Agricultural Engineers 20:271-275.

Campbell, R. B., D. C. Reicosky, and C. W. Doty. 1974. Physical properties and tillage of Paleudults in the Southeastern Coastal Plains. Journal of Soil and Water Conservation 29:220-224.

Chaney, K., and R. S. Swift. 1986. The influence of organic matter on aggregate stability in some British soils. Journal of Soil Science 35:223-230.

Chapman, S. J., and J. M. Lynch. 1984. A note on the formation of microbial polysaccharide from wheat straw decomposed in the absence of soil. Journal of Applied Bacteriology 56:337-342.

Chapman, S. J., and J. M. Lynch. 1985. Polysaccharide synthesis by capsular microorganisms in coculture with cellulolytic fungi on straw and stabilization of soil aggregates. Biology and Fertilization of Soils 1:161-166.

Cregan, P. B., and H. H. Keyser. 1986. Host restriction of nodulation by Bradyrhizobium japonicum strain USDA 123 in soybean. Crop Science 26:911-916.

Dobereiner, J. 1961. Nitrogen-fixing bacteria of the genus Beijerinckia derx in the rhizosphere of sugar cane. Plant and Soil 3:211-217

Dobereiner, J., and J. M. Day. 1976. Associative symbiosis and free-living systems. In Proceedings of the first international symposium on N_2 fixation, ed. W. E. Newton and C. J. Nyman, 518-538. Pullman: Washington State University Press.

Dobereiner, J., J. M. Day, and P. J. Dart. 1972. Nitrogenase activity and oxygen sensitivity of the Paspalum notatum-Azotobacter paspali association. Journal of General Microbiology 71:103-116.

Dunigan, E. P., P. K. Bollich, R. L. Hutchinson, P. M. Hicks, F. C. Zaunbrecher, S. G. Scott, and R. P. Mowers. 1984. Introduction and survival of an inoculant strain of Rhizobium japonicum in soil. Agronomy Journal 76:463-466.

Elliott, L. F., and J. M. Lynch. 1984. The effect of available carbon and nitrogen in straw on soil and ash aggregation and acetic acid production. Plant and Soil 78:335-343.

Frankenberger, W. T., Jr., and M. Poth. 1987. Biosynthesis of indole-3-acetic acid by the pine ecotomycorrhizal fungus Pisolithus tinctorius. Applied and Environmental Microbiology 53:2908-2913.

Fuhrmann, J., C. B. Davey, and A. G. Wollum II. 1986. Desiccation tolerance of clover rhizobia in sterile soils. Soil Science Society of America Journal 50:639- 644.

Gaskins, M. H., and J. C. Carter. 1975. Nitrogenase activity: A review and evaluation of assay methods. Soil Crop Science Society of Florida 35:10-16.

Giddens, J. E., E. P. Dunigan, and R. W. Weaver. 1982. Legume inoculation in the Southeastern U.S.A. University of Georgia, Southern Cooperative Series Bulletin no. 283. pp. 35.

Harper, S. H. T., and J. M. Lynch. 1984. Nitrogen fixation by cellulolytic communities at aerobic-anaerobic interfaces in straw. Journal of Applied Bacteriology 57:131-137.

Hashem, F. M., and J. S. Angle. 1988. Rhizobiophage effects on Bradyrhizobium japonicum, nodulation and soybean growth. Soil Biology and Biochemistry 20:69-73.

Heichel, G. H. 1987. Legumes as a source of nitrogen in conservation tillage systems. In Proceedings of the national conference, the role of legumes in conservation tillage systems ed. J. F. Power. Soil Conservation Society of America, University of Georgia, Athens, April 27-29.

Hiltbold, A. E., R. M. Patterson, and R. B. Reed. 1985. Soil populations of Rhizobium japonicum in a cotton corn-soybean rotation. Soil Science Society of America Journal 49:343-348.

Hubbell, D. H., and M. H. Gaskins. 1984. Associative N_2 fixation with Azospirillum. In Biological nitrogen fixation. ed. Martin Alexander, 201-224. New York: Plenum Publishing Corporation.

Hunt, P. G., T. Kaneshiro, and T. A. Matheny. 1987. Influence of IAA-producing B. japonicum on growth and yield of soybean. Agronomy Abstracts p. 185.

Hunt, P. G., M. J. Kasperbauer, and T. A. Matheny. 1987. Nodule development in a split-root system in response to red and far-red light treatment of soybean shoots. Crop Science 27:973-976.

Hunt, P. G., M. J. Kasperbauer, and T. A. Matheny. 1989. Soybean seedling growth responses to light reflected from different colored soil surfaces. Crop Science 29:130-133.

Hunt, P. G., M. J. Kasperbauer, and T. A. Matheny. 1990. Influence of Bradyrhizobium japonicum strain on nodulation of soybean with varying far-red/red ratios in canopy light. Crop Science (In press).

Hunt, P. G., T. A. Matheny, and A. G. Wollum II. 1985. Rhizobium japonicum nodular occupancy, nitrogen accumulation, and yield for determinate soybean under conservation and conventional tillage. Agronomy Journal 77:579-584.

Hunt, P. G., T. A. Matheny, and A. G. Wollum II. 1988. Yield and N accumulation response of late-season determinate soybean to irrigation and inoculation with various strains of Bradyrhizobium japonicum. Communication in Soil Science and Plant Analysis 19:1601-1612.

Hunt, P. G., T. A. Matheny, A. G. Wollum II, D. C. Reicosky, R. E. Sojka, and R. B. Campbell. 1983. Effect of irrigation and Rhizobium japonicum strain 110 upon yield and nitrogen accumulation and distribution in determinate soybeans. Communication in Soil Science and Plant Analysis 14:223-238.

Hunt, P. G., R. E. Sojka, T. A. Matheny, and A. G. Wollum II. 1985. Soybean response to R. japonicum strain, row orientation and irrigation. Agronomy Journal 77:720-725.

Hunt, P. G., A. G. Wollum II, and T. A. Matheny. 1981. Effects of soil water on R. japonicum infection, nitrogen accumulation and yield in Bragg soybeans. Agronomy Journal 73:501-505.

Jordan, D. C. 1982. Transfer of Rhizobium japonicum Buchanan 1980 to Bradyrhizobium gen. nov., a genus of slow-growing root nodule bacteria from leguminous plants. International Journal of Systemic Bacteriology 32:136-139.

Kaneshiro, T., and W. F. Kwolek. 1985. Stimulated nodulation of soybeans by Rhizobium japonicum mutant (B-14075) that catabolizes the conversion of tryptophan to indole-3-acetic acid. Plant Science 42:141-146.

Kapulnik, Y., J. Kigel, Y. Okon, I. Nur, and Y. Henis. 1981a. Effect of Azospirillum inoculation on some growth parameters and N-content of wheat, sorghum and panicum. Plant and Soil 61:65-70.

Kapulnik, Y., Y. Okon, J. Kigel, I. Nur, and Y. Henis. 1981b. Effects of temperature, nitrogen fertilization, and plant age on nitrogen fixation by Setaria italica inoculated with Azospirillum brasilense (strain cd). Plant Physiology 68:340-343.

Karlen, D. L., and P. G. Hunt. 1985. Copper, nitrogen, and Rhizobium japonicum relationships in determinate soybean. Journal of Plant Nutrition 8:395-404.

Kasperbauer, M. J., and P. G. Hunt. 1987. Soil color and surface residue effects on seedling light environment. Plant and Soil 97:295-298.

Kasperbauer, M. J., and P. G. Hunt. 1988. Biological and photometric measurement of light transmission through soils of various colors. Botanical Gazette 149:361-364.

Kasperbauer, M. J., P. G. Hunt, and R. E. Sojka. 1984. Photosynthate partitioning and nodule formation in soybean plants that received red or far-red light at the end of the photosynthetic period. Physiologia Plantarum 61:549-554.

Kavimandan, S. K. 1986. Influence of rhizobial inoculation on yield of wheat (Triticum aestivum L.). Plant and Soil 95:297-300.

Keyser, H. H., and D. N. Munns. 1979a. Effects of calcium, manganese and aluminum on growth of rhizobia in acid media. Soil Science Society of America Journal 43:500- 503.

Keyser, H. H., and D. N. Munns. 1979b. Tolerance of rhizobia to acidity, aluminum, and phosphate. Soil Science Society of America Journal 43:519-523.

Keyser, H. H., B. B. Bohlool, T. S. Hu, and D. F. Weber. 1982. Fast-growing rhizobia isolated from root nodules of soybean. Science 215:1631-1632.

Kosslak, R. M., and B. B. Bohlool. 1983. Prevalence of Azospirillum spp. in the rhizosphere of tropical plants. Canadian Journal of Microbiology 29:649-652.

Kvien, C. S., and G. E. Ham. 1985. Effect of soil temperature and inoculum rate on the recovery of three introduced strains of Rhizobium japonicum. Agronomy Journal 77:484-489.

Lin, W., Y. Okon, and R. W. F. Hardy. 1983. Enhanced mineral uptake by Zea mays and Sorghum bicolor roots inoculated with Azospirillum brasilense. Applied and Environmental Microbiology 45:1775-1779.

Lynch, J. M., J. H. Slater, J. A. Bennett, and S. H. T. Harper. 1981. Cellulase activities of some aerobic micro-organisms isolated from soil. Journal of General Microbiology 127:231-236.

Mahler, R. L., and A. G. Wollum II. 1981. The influence of soil water potential and soil texture on the survival of Rhizobium japonicum and Rhizobium leguminosarum isolates in the soil. Soil Science Society of America Journal 45:761-766.

Materon, L. A., and R. W. Weaver. 1984. Survival of Rhizobium trifolii on toxic and non-toxic arrowleaf clover seeds. Soil Biology and Biochemistry 16:533- 535.

Matheny, T. A., and P. G. Hunt. 1983. Effects of irrigation on accumulation of soil and symbiotically fixed N by soybean grown on a Norfolk loamy sand. Agronomy Journal 75:719-722.

Mathis, J. N., L. D. Kuykendall, and G. H. Elkan. 1986. Restriction endonuclease and nif homology patterns of Bradyrhizobium japonicum USDA 110 derivatives with and without nitrogen fixation competence. Applied and Environmental Microbiology 51:477-480.

Meshram, S. U., and S. T. Shende. 1982. Total nitrogen uptake by maize with Azotobacter inoculation. Plant and Soil 69:275-279.

Moawad, H. A., and B. B. Bohlool. 1984. Competition among Rhizobium spp. for nodulation of Leucaena leucocephala in two tropical soils. Applied and Environmental Microbiology 48:5-9.

Moawad, H. A., W. R. Ellis, and E. L. Schmidt. 1984. Rhizosphere response as a factor in competition among three serogroups of indigenous Rhizobium japonicum for nodulation of field-grown soybeans. Applied and Environmental Microbiology 47:607-612.

Morris, D. R., and R. W. Weaver. 1983. Mobilization of ^{15}N from soybean leaves as influenced by rhizobial strains. Crop Science 23:1111-1114.

Munevar, F., and A. G. Wollum II. 1981. Effect of high root temperature and Rhizobium strain on nodulation, nitrogen fixation, and growth of soybeans. Soil Science Society of America Journal 45:1113-1120.

Newman, E. I. 1985. The rhizosphere: Carbon sources and microbial populations. Special Publication Series of the British Ecological Society 4:107-121.

Neyra, C. A., and J. Dobereiner. 1977. Nitrogen fixation in grasses. Advances in Agronomy 29:1-51.

O'Hara, G. W., M. R. Davey, and J. A. Lucas. 1981. Effect of inoculation of Zea mays with Azospirillum brasilense strains under temperate conditions. Canadian Journal of Microbiology 27:871-877.

O'Hara, G. W., M. J. Dilworth, N. Boonkerd, and P. Parkpian. 1988. Iron-deficiency specifically limits nodule development in peanut inoculated with Bradyrhizobium sp. New Phytology 108:51-57.

Osa-Afiana, L. O., and M. Alexander. 1982. Differences among cowpea rhizobia in tolerance to high temperature and desiccation in soil. Applied and Environmental Microbiology 43:435-439.

Pena-Cabriales, J. J., and M. Alexander. 1979. Survival of Rhizobium in soils undergoing drying. Soil Science Society of America Journal 43:962-966.

Piha, M. I., and D. N. Munns. 1987a. Nitrogen fixation capacity of field-grown bean compared to other grain legumes. Agronomy Journal 79:690-696.

Piha, M. I., and D. N. Munns. 1987b. Nitrogen fixation potential of beans (Phaseolus vulgaris L.) compared with other grain legumes under controlled conditions. Plant and Soil 98:169-182.

Rai, R. 1985a. Manganese-resistance mutants of Azospirillum brasilense: Their growth and relative efficiency in associative nitrogen fixation with cheena (Panicum miliaceum) in acid soil. Journal of General Applied Microbiology 31:211-219.

Rai, R. 1985b. Studies on associative nitrogen fixation by antibiotic-resistant mutants of Azospirillum brasilense and their interaction with lentil (Lens culinaris) rhizobium strains in calcareous soil. Journal of Agricultural Science 104:207-215.

Rai, R. 1985c. Studies on nitrogen fixation by antibiotic resistant mutants of Azospirillum brasilense and their interaction with cheena (Panicum miliaceum) genotypes in calcareous soil. Journal of Agricultural Science 105:261-270.

Rai, R. 1987. Chemotaxis of salt-tolerant and sensitive Rhizobium strains to root exudates of lentil (Lens culinaris L.) genotypes and symbiotic N-fixation, proline content and grain yield in saline calcareous soil. Journal of Agricultural Science 108:25-37.

Rai, R, S. K. T. Nasar, S. J. Singh, and V. Prasad. 1985. Interactions between Rhizobium strains and lentil (Lens culinaris Linn.) genotypes under salt stress. Journal of Agricultural Science 104:199-205.

Rai, S. N., and A. C. Gaur. 1982. Nitrogen fixation by Azospirillum spp. and effect of Azospirillum lipoferum on the yield and N-uptake of wheat crop. Plant and Soil 69:233-237.

Rao, V. R., A. Ayanaba, A.R.J. Eaglesham, and G. Thottappilly. 1985. Effects of Rhizobium inoculation on field-grown soybeans in Western Nigeria and assessment of inoculum persistence during a two-year fallow. Tropical Agriculture 62:125-130.

Reynders, L., and K. Vlassak. 1982. Use of Azospirillum brasilense as biofertilizer in intensive wheat cropping. Plant and Soil 66:217-233.

Riley, I. T., and M. J. Dilworth. 1985. Cobalt status and its effects on soil populations of Rhizobium lupini, rhizosphere colonization and nodule initiation. Soil Biology and Biochemistry 17:81-85.

Rupela, O. P., B. Toomsan, S. Mittal, P. J. Dart, and J. A. Thompson. 1987. Chickpea Rhizobium populations: Survey of influence of season, soil depth and cropping pattern. 1987. Soil Biology and Biochemistry 19:247-252.

Sadowsky, M. J., H. H. Keyser, and B. B. Bohlool. 1983. Biochemical characterization of fast- and slow-growing rhizobia that nodulate soybeans. International Journal of Systematic Bacteriology 33(4):716-722.

Sarig, S. Y. Kapulnik, I. Nur, and Y. Okon. 1984. Response of non-irrigated Sorghum bicolor to Azospirillum inoculation. Experimental Agriculture 20:59-66.

Smith, R. L., J. H. Bouton, S. C. Schank, K. H. Quesenberry, M. E. Tyler, J. R. Milam, M. H. Gaskins, and R. C. Littell. 1976. Nitrogen fixation in grasses inoculated with Spirillum lipoferum. Science 193:1003-1005.

Thurlow, D. L., and A. E. Hiltbold. 1985. Dinitrogen fixation by soybeans in Alabama. Agronomy Journal 77:432-436.

Tien, T. M., H. G. Diem, M. H. Gaskins, and D. H. Hubbell. 1981. Polygalacturonic acid transeliminase production by *Azospirillum* species. *Canadian Journal of Microbiology* 27:426-431.

Tien, T. M., M. H. Gaskins, and D. H. Hubbell. 1979. Plant growth substances produced by *Azospirillum brasilense* and their effect on the growth of Pearl millet (*Pennisetum americanum* L.). *Applied and Environmental Microbiology* 37:1016-1024.

Trinick, M. J., R. L. Rhodes, and J. H. Galbraith. 1983. Competition between fast- and slow-growing tropical legume rhizobia for nodulation of *Vigna unguiculata*. *Plant and Soil* 73:105-115.

Trotman, A. A., and R. W. Weaver. 1986. Number and effectiveness of cowpea rhizobia in soils of Guyana. *Tropical Agriculture* 63:129-132.

van Berkum, P., and B. B. Bohlool. 1980. Evaluation of nitrogen fixation by bacteria in association with roots of tropical grasses. *Microbiology Review* 44:491-517.

Viteri, S. E., and E. L. Schmidt. 1987. Ecology of indigenous soil rhizobia: Response of *Bradyrhizobium japonicum* to readily available substrates. *Applied and Environmental Microbiology* 53:1872-1875.

von Bulow, J. F. W., and J. Dobereiner. 1975. Potential for nitrogen fixation in maize genotypes in Brazil. *Proceedings of the National Academy of Science, USA* 72:2389-2393.

Weaver, R. W., and L. R. Frederick. 1974. Effect of inoculum rate on competitive nodulation of *Glycine max* L. Merrill. II. Field studies. *Agronomy Journal* 66:233-236.

Weaver, R. W., L. R. Frederick, and L. C. Dumenil. 1972. Effect of soybean cropping and soil properties on numbers of *Rhizobium japonicum* in Iowa soils. *Soil Science* 114:137-141.

Welch, L. F., L. V. Boone, C. G. Chambliss, A. T. Christiansen, D. L. Mulvaney, M. G. Oldham, and J. W. Pendleton. 1973. Soybean yields with direct and residual nitrogen fertilization. *Agronomy Journal* 65:547-550.

Whelan, A. M., and M. Alexander. 1986. Effects of low pH and high Al, Mn and Fe levels on the survival of *Rhizobium trifolii* and the nodulation of subterranean clover. *Plant and Soil* 92:363-371.

Whipps, J. M., and J. M. Lynch. 1983. Substrate flow and utilization in the rhizosphere of cereals. *New Phytology* 95:605-623.

Williams, L. E., and D. A. Phillips. 1983. Increased soybean productivity with a *Rhizobium japonicum* mutant. *Crop Science* 23:246-250.

Wollum, A. G., II, and D. K. Cassel. 1984. Spatial variability of *Rhizobium japonicum* in two North Carolina soils. *Soil Science Society of America Journal* 48:1082-1086.

Young, N. R., D. M. Hughes, and L. R. Mytton. 1986. The response of white clover to different strains of *Rhizobium trifolii* in hill land reseeding: A second trail. *Plant and Soil* 94:277-284.

David M. Sylvia

6 Distribution, Structure, and Function of External Hyphae of Vesicular-Arbuscular Mycorrhizal Fungi

The most widespread fungi in the rhizospheres of An-giosperms are Zygomycetes that form symbiotic associations with plant roots —the vesicular-arbuscular mycorrhizal (VAM) fungi (Harley and Harley 1987; Trappe 1987). The biology of VAM fungi has been reviewed by many authors, including Hayman (1978); Mosse, Stribley, and LeTacon (1981); Gianinazzi-Pearson and Diem (1982); Moser and Haselwandter (1983); Harley and Smith (1983); and Jeffries (1987). Root colonization by VAM fungi often leads to improved plant growth in nutrient-poor soils, primarily because of increased uptake of phosphorus and other poorly mobile ions. Furthermore, mycorrhizal plants often are more tolerant of water stress, pathogens, salt, and heavy metals than nonmycorrhizal plants. These responses are thought to be related directly to improved phosphorus nutrition, however (Graham 1987).

Root colonization by VAM fungi has both an internal and an external phase. The development of infections with-in the root has been studied in detail (Sanders and Sheikh 1983; Smith and Walker 1981; Bonfante-Fasolo 1984). Prima-ry infections are initiated from appressoria, which form on the root surface. Inside the root, hyphae branch and grow between the cells of the outer cortex to the inner cortex, where intracellular penetration occurs and arbuscules form. Secondary infections are often initiated from hyphae that proliferate on the outside of roots. Vesicles usually form later, often as arbuscules begin to senesce (Douds and Chaney 1982).

The external phase of VAM fungal growth has not re-ceived adequate study. Even though the hyphae that radiate into the soil matrix from the root are the functional or-gans for nutrient uptake and translocation, few researchers have provided quantitative data on their growth and distribution. Read (1984) presented an ecophysiological perspective on these absorptive structures. New research reports and the critical need for further evaluations of the external phase of VAM fungal development, however, warrant additional appraisal of this topic. The objectives of this chapter are to discuss current knowledge of the distribution, structure, and function of the external hyphae of VAM fungi; to evaluate current techniques for assessing development of VAM hyphae in soil; and to suggest research needs in this critical area of mycorrhiza biology.

DISTRIBUTION OF EXTERNAL HYPHAE

Quantity of Hyphae Produced in Soil

Fungal hyphae are abundant in soil. Kjøller and Struwe (1982) summarized data from three ecosystems— tundra, grassland, and woodland. They reported that total hyphal lengths in the 20 cm at the surface ranged from 100 to 10,000 m/g. The proportion in native soils of total hyphae that belong to symbiotic fungi is not known because there are no reliable methods to distinguish VAM fungal hyphae from other hyphae in the soil. Data from pot studies, however, suggest that there may be large variation in the VAM fungal contribution to total fungal biomass. Abbott, Robson, and DeBoer (1984) and Sylvia (1988) found only small increases in total hyphae in pots inoculated with VAM fungi compared with noninoculated pots, whereas Schubert et al. (1987) reported an increase in hyphal length of more than 800 percent following VAM fungal inoculation.

Estimates of the amount of VAM fungal hyphae in soil vary greatly. Besides the inherent differences among experimental systems, a major problem in summarizing data on external hyphae is that different components of the external phase have been measured. Some authors measured only hyphae attached to roots (here called rhizoplane hyphae), others measured hyphae extracted from soil (here called soil hyphae), and still others reported both these values, combined or separated. The remainder of this section provides estimates from the extant literature for maximum values of external hyphae.

Data on external hyphae are often reported on a root-length basis. Maximum reported values for rhizoplane hyphae (cm/cm of root) are 6 (Diem et al. 1986) and 3

(Sylvia 1988). Reported maximum lengths for soil hyphae (m/cm of root) are 592 (Sylvia 1986), 30 (Abbott and Robson 1985b), 13 (Abbott et al. 1984), and less than 1 (Sanders and Tinker 1973; Tisdall and Oades 1979). The unusually high value reported by Sylvia (1986) was from a native sand dune where rooting density was low. On a soil mass basis, maximum reported values for soil hyphae (m/g of soil) are 26 (Abbott and Robson 1985b), 20 (Tisdall and Oades 1979), 16 (Bethlenfalvay and Ames 1987), 15 (Abbott et al. 1984), 12 (Sylvia 1986), 6 (Abbott and Robson 1985a), 4 (Sylvia 1988), and less than 1 (Schubert et al. 1987). In rhizosphere soil, Allen and Allen (1986) recorded values up to 54 m/g.

The biomass of external hyphae has also been estimated by several workers. On a soil basis, Nicolson and Johnston (1979) recorded up to 2.5 g/l of dune sand (about 1.8 mg/g), but their sample included an unknown amount of debris. Bethlenfalvay and Ames (1987) found a hyphal mass of 40 μg/g of soil using a chitin assay in a pot culture study. On a root basis, Bethlenfalvay, Brown, and Pacovsky (1982) reported 47 mg/g of root, but this estimate included spores. Sanders et al. (1977) found 3.6 μg/cm of infected root length. Finally, Kucey and Paul (1982) estimated that attached hyphae represented two percent of the total root biomass, but they failed to report actual values.

Most researchers who attempt to quantify external hyphae of VAM fungi use methods that do not differentiate between active and inactive or dead hyphae. Since phosphorus uptake and translocation by fungi is an active process (Beever and Burns 1980), quantification of active hyphae is important. To my knowledge, only two studies report the development of active external hyphae in pot cultures over time. Schubert et al. (1987) reported a maximum of 15.4 cm/g of soil forty-eight days after inoculation, followed thereafter by a rapid decline. The proportion of active hyphae in their cultures declined steadily with time (see Figure 6.1). Sylvia (1988) found more than 100 cm/g of soil sixty-three days after inoculation and reported that the proportion of active hyphae increased with time (Figure 6.2). Discrepancies in these data are attributable to different growth environments (Schubert et al. used a low-light growth chamber, whereas Sylvia used a high-light glasshouse); host and fungal species; and levels of background hyphae in the substrate. One tentative conclusion from these data is that fungal turnover rates are much more

rapid under low-light than under high-light growth conditions.

Factors Affecting Distribution of External Hyphae

Little is known about the effects of various biotic
and abiotic factors on the distribution of external hyphae
of VAM fungi. Probably many of the chemical, physical, and
biotic properties of the soil that influence plant response
to VAM fungal colonization act directly on the external
phase of the symbiosis. The few known interactions are discussed below; however, the brevity of the data emphasizes
the need for more study in this area.

Phosphorus concentration is known to affect the function of mycorrhizae. Same, Robson, and Abbott (1983) found
that small additions of phosphorus to phosphorus-deficient
plants stimulated the internal phase of root colonization,
whereas higher phosphorus applications reduced the percent-

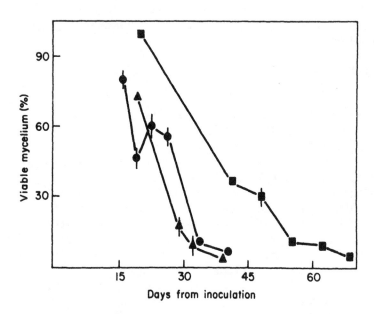

Figure 6.1. Proportion of viable to total external hyphal
length of Glomus clarum inoculated on Trifolium repens in
three experiments.

Source: Schubert et al. 1987. Reprinted by permission.

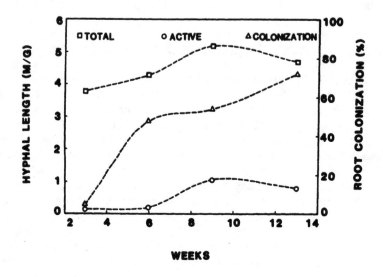

Figure 6.2.   Total and active hyphae and root colonization in pot cultures of <u>Glomus spp.</u> grown with <u>Paspalum notatum</u> for 13 weeks.

<u>Source</u>:   Sylvia 1988.

age of root length colonized. Abbott, Robson, and DeBoer (1984) reported that external hyphae had a similar response to phosphorus (Figure 6.3); small additions stimulated, whereas large additions suppressed, hyphal growth. Schwab, Menge, and Leonard (1983) also found that high phosphorus application reduced the proliferation of external hyphae.

Other chemical and physical properties of soil that have been shown to affect external hyphae are pH, copper concentration, organic matter content, and water. Mycorrhizal fungi respond differently to changes in soil pH. Abbott and Robson (1985a) found one <u>Glomus</u> isolate to be infective over a wide pH range (5.3 to 7.5), although another isolate was infective only at the highest pH level. The lack of spread of the pH-sensitive isolate was associated with the inability of the fungus to produce external hyphae at a low pH. The effects of high levels of copper in soil on the development of VAM fungi on citrus was studied by Graham, Timmer, and Fardelmann (1986); copper-induced reduction in phosphorus uptake by VAM plants was closely related

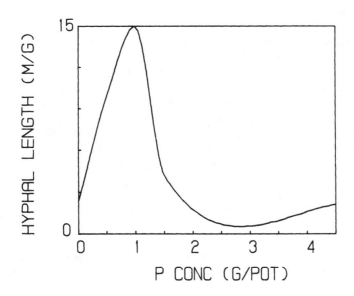

Figure 6.3. Effect of applied phosphorus on length of VAM hyphae in soil. Values were obtained by subtracting the total length of hyphae in noninoculated pots from that in inoculated pots.

Source: Redrawn from Abbott, Robson, and DeBoer 1984. Reprinted by permission.

to inhibition of rhizosphere hyphae. St. John, Coleman, and Reid (1983) reported that organic particles in sand increased growth of external hyphae and concluded that organic matter stimulated hyphal branching. Finally, Bethlenfalvay and Ames (1987) found that external hyphal lengths and biomass increased with water stress.

Only a few studies explore the effects of soil microflora and fauna on external hyphae. St. John, Coleman, and Reid (1983) inoculated plant roots with a mixture of soil microorganisms and found that this treatment inhibited hyphal growth. The researchers speculated that actinomycetes were involved in the inhibition. Warnock, Fitter, and Usher (1982) reported that collembola in the soil graze on VAM fungal hyphae. Fitter (1985, 1986) suggested that such grazing severely limits the effectiveness of VAM fungi in native soils.

The VAM fungi vary considerably in their innate capacity to produce external hyphae. Abbott and Robson (1985b) found that <u>Glomus fasciculatum</u> produced less external hyphae from <u>Trifolium subterraneum</u>-colonized roots than did three other VAM fungi (Figure 6.4). In their study, the production of external hyphae was not related to percentage of root colonization but was associated with plant growth enhancement. Graham, Linderman, and Menge (1982) reported similar results with citrus. Plants became well colonized by seven isolates of VAM fungi, but only those fungi that produced a large amount of external hyphae, estimated by weighing soil attached to roots, produced a concomitant growth response (Figure 6.5).
Sanders et al. (1977) reported that <u>Glomus microcarpum</u> produced fewer external hyphae than three other endophytes tested, but in their study, poor development of external hyphae was associated with a slow rate of root colonization and no plant growth response. These studies indicate the critical need to clarify the relationships among root colonization, production of external hyphae, and plant growth response.

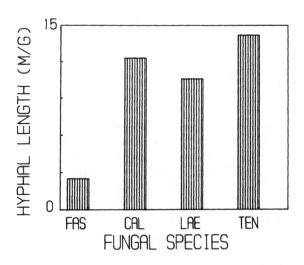

Figure 6.4. Production of hyphae of <u>Glomus fasciculatum</u> (FAS), <u>G. tenue</u> (TEN), <u>Gigaspora calospora</u> (CAL), and <u>Acaulospora laevis</u> (LEA) in soil.

<u>Source</u>: Redrawn from Abbott and Robson 1985b. Reprinted by permission.

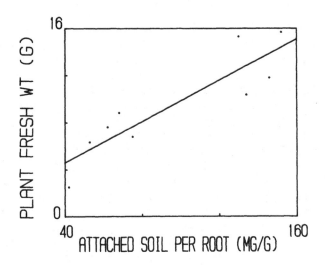

Figure 6.5.   Relationship between plant growth and production of external hyphae by VAM fungi.

<u>Source</u>:   Redrawn from Graham, Linderman, and Menge 1982.
Reprinted by permission.

Plant factors also have an effect on the production of
external hyphae.   Plant genotype affects root colonization
(Heckman and Angle 1987) and would be expected to affect
external hyphae.   Tisdall and Oades (1979) found that ryegrass supported more hyphae in the soil than white clover,
but this may be due to the different rooting densities of
these plants. Additional studies on the effects of plant
genotype on hyphal development are needed.   Since VAM fungi
derive their carbon from a host plant, treatments that
reduce photosynthetic capacity will also reduce mycorrhiza
development.   For example, shading and defoliation have
been shown to reduce root colonization and vesicle formation (Bethlenfalvay and Pacovsky 1983; Same, Robson, and
Abbott 1983).   These same treatments probably affect external hyphae, although no published research has specifically
addressed these interactions.

MORPHOLOGY AND DEVELOPMENT OF EXTERNAL HYPHAE

The dimorphic nature of external hyphae of VAM fungi has been known for many years.[c] Mosse (1959) and Nicolson (1959) described coarse, thick-walled hyphae having diameters of more than 20 μm and fine, thin-walled hyphae with diameters ranging from 2 to 10 μm. Mosse (1959) found the highest proportion (more than 70 percent) of the hyphae in her samples were of the coarse type. Coarse hyphae are characteristic of VAM fungi; they are usually aseptate and have distinct angular projections, as described by Butler (1939). Fine hyphae appear to be short-lived, are often septate, and cannot be readily distinguished from other hyphae in the soil. Nicolson (1959) observed that fine hyphae develop in two ways: by direct development as lateral branches from coarse hyphae, and by repeated branching of coarse hyphae giving progressively finer filaments. Powell (1976) also observed repeated branching of coarse VAM fungal hyphae resulting in fanlike structures near the root. He presumed these to be pre-infection structures. Due to the various branching patterns, Mosse (1959) was correct to state that intermediate types of hyphae may also be observed.

Several researchers have used hyphal diameter to distinguish VAM fungi from other fungi in the soil. For example, Graham, Timmer, and Fardelman (1986) and Bethlenfalvay and Ames (1987) considered those hyphae more than 5 μm in diameter as VAM fungal hyphae. Abbott and Robson (1985b), however, found that most hyphae in VAM fungal-inoculated pot cultures had diameters between 1 and 5 μm. They concluded that it was not possible to distinguish hyphae of VAM fungi from nonmycorrhizal fungi by morphological or staining criteria. Serological techniques, however, may permit specific detection of mycorrhizal hyphae in soil (Kough et al. 1986).

The apparent discrepancy between Abbott and Robson's finding (most hyphae less than 5 μm in diameter) and Mosse's observation (more than 70 percent coarse hyphae) may be related to culture age and growth environment. The former observations were from short-term pot culture studies, while the latter were from nursery-grown trees. Fine hyphae are short-lived and ephemeral, while coarse hyphae are long-lived and recalcitrant. It may be that fine hyphae predominate in young infections and that over time a greater proportion of the fungal mass achieves a large diameter. Unfortunately, actual turnover rates for these structures are unknown. I consider clarification of the growth dynamics of external hyphae most important to further understanding of their function in the soil.

Sutton (1973) described three phases in the development of VAM fungal infections (a lag phase, followed by a phase of exponential development, and finally a constancy or plateau phase). More recently, the development of VAM fungal infections in roots was reviewed by Bowen (1987). Only a few researchers have quantified the development of the external hyphae in relation to the internal phases of development, however, and these data are not always in agreement. There are examples of internal hyphae developing before external hyphae, of a constant relationship between internal and external hyphal development, and of external hyphae developing before internal hyphae. Sylvia (1988) and Diem et al. (1986) found that extensive internal infections occurred before development of the external hyphae. In contrast, Bethlenfalvay, Brown, and Pacovsky (1982) reported that the ratio of external to internal hyphae decreased with time, indicating that the external phase initially developed more rapidly than the internal phase. Schubert et al. (1987) reported that the viable length of external hyphae decreased with time; however, data on internal infection were not presented, limiting the application of these data to this question. Finally, Sanders et al. (1977) found a constant relationship between external hyphae production and root infection. Many more experiments, conducted over time, are needed to clarify the relationship between the development of internal and external hyphae. It is apparent that VAM fungi have different modes of root/soil colonization (Scheltema et al. 1987; Wilson 1984), and selection of effective VAM fungal isolates may be facilitated by a better understanding of this process.

FUNCTION OF HYPHAE IN SOIL

Nutrient and Water Uptake and Translocation

Improved uptake of poorly mobile ions from the soil is the major benefit of VAM fungi to plants. The structures involved in nutrient uptake are clearly the external hyphae. The mechanisms of nutrient uptake by hyphae have been reviewed by several authors (Tinker and Gildon 1983; Cooper 1984). I will point out some gaps in our knowledge and suggest areas of additional research.

Rhodes and Gerdemann (1975) showed unequivocally that external hyphae of VAM fungi take up phosphorus from the soil and translocate it some distance to the plant (Figure 6.6). More recently, Alexander, Alexander, and Hadley (1984) found transport to occur over distances of 9 cm with an orchid mycorrhiza. These studies, however, were conducted in model systems using autoclaved soil. Fitter (1985) suggested that under natural conditions, hyphal integrity may be compromised and intact hyphal lengths for transport may be greatly reduced. The work of Owusu-Bennoah and Wild (1979) supports Fitter's contention; they found that depletion zones around mycorrhizae in sterile soil were greater than those around mycorrhizae in nonsterile soil. Further evaluation of nutrient uptake and translocation by mycorrhizae in natural soils is needed. Use of techniques such as sectioning of intact soil cores (Skinner and Bowen 1974) and autoradiography (Owusu-Bennoah and Wild 1979) should greatly increase our understanding of the function of

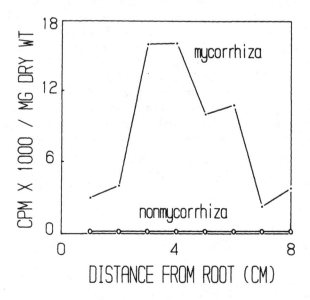

Figure 6.6. ^{32}P present in root segments from mycorrhizal and nonmycorrhizal onions as a function of distance from tracer injection point to root surface.

Source: Rhodes and Gerdemann 1975. Reprinted by permission.

hyphae in natural systems. Little is known about the zone of nutrient uptake around external hyphae. Rhodes and Gerdemann (1975) observed that maximum phosphorus uptake by mycorrhizal roots occurred several centimeters away from the root surface, in the zone of young, thin-walled hyphae (Figure 6.6). Furthermore, hyphae do not have uniform metabolic activity (Schubert et al. 1987). The implications of this for nutrient uptake and translocation are not known. Uptake of phosphorus requires metabolically active hyphae (Cooper and Tinker 1981); however, translocation through inactive zones needs to be evaluated.

Nelson (1987) recently reviewed the water relations of VAM plants and concluded that most changes attributed to VAM fungi were secondary responses due to improved plant nutrition. Indeed, improved nutrient and water uptake by VAM plants are closely related phenomena, since nutrients are available to the plant only in the soil solution. Under conditions of water stress, movement of nutrients in the root zone is restricted, limiting uptake of poorly mobile ions (Viets 1972). Hardie and Leyton (1981), however, demonstrated improved hydraulic conductivities of VAM plants versus nonmycorrhizal plants over a range of applied phosphorus (Figure 6.7). These authors concluded that improved conductivities were mainly due to hyphal growth in the soil. In contrast, Graham and Syvertsen (1984) found that water flow to roots via hyphae could not account for greater water uptake by mycorrhizal roots. Further evaluation of this phenomenon, especially under field conditions, is needed.

Nutrient Transfer Among Plants

Intra- and interspecific nutrient transfers via mycorrhizal hyphae are known to occur (Read 1984). For VAM systems, transfers of carbon (Hirrel and Gerdemann 1979; Francis and Read 1984), nitrogen (van Kessel et al. 1985; Francis et al. 1986), and phosphorus (Heap and Newman 1980; Chiariello et al. 1982; Whittingham and Read 1982; Francis et al. 1986) have been demonstrated.

The extent and implications of these interconnections are not known. If significant movements of nutrients are occurring among plants, many studies on nutrient cycling and plant competition will need to be reexamined. In a recent microcosm study, Grime et al. (1987) found that the presence of VAM fungi resulted in increased species diver-

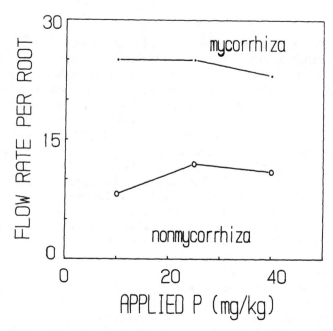

Figure 6.7. Water conductivity, units for flow rate are cm³ sec⁻¹ cm⁻¹ x 10⁻⁸, of roots of mycorrhizal and nonmycorrhizal red clover plants grown at three phosphorus levels.

Source: Hardie and Leyton 1981. Reprinted by permission.

sity. They concluded that this was due to transfer of nutrients and carbon to small-seeded species which would not have survived without the hyphal links. Whether significant amounts of nutrients are directly transferred among plants by hyphal connections, however, is not yet resolved (Newman and Ritz 1986).

Soil Aggregation

As stated above, meters of VAM hyphae are often observed per gram of soil. It would seem that this mass of hyphae would act to bind soil particles together, and several studies suggest that this does happen. Koske, Sutton, and Sheppard (1975) found extensive hyphal networks of endogonaceous fungi binding sand among plant roots in sand dunes. Sutton and Sheppard (1976) followed up these field observations with an experiment in which benomyl was applied to some treatments to prevent mycorrhizal develop-

ment.   They reported that reduced mycorrhiza resulted in
reduced aggregation of soil particles.   Their conclusion
was that the major mechanism linking sand grains in aggregates was the binding action of the extensive and persistent Glomus hyphae.   Tisdall and Oades (1979) found a good
correlation between the amount of hyphae produced in soil
and the production of water-stable aggregates (Figure 6.8).
    Lynch and Bragg (1985) expressed some reservation
about the direct role of fungi in soil stabilization.   They
felt that the role of hyphae may be indirect, serving as a
substrate for other polysaccharide-producing microorganisms.   Polysaccarides have been observed on the surface of
VAM fungal hyphae (Tisdall and Oades 1979).   A tenable
model for enhanced soil stabilization by VAM fungi is that
the external hyphae, along with plant roots, produce a
framework for aggregation, while bacterial polysaccarides
cement the soil particles together.

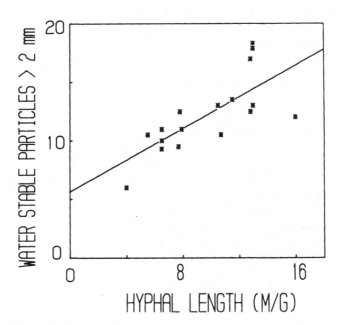

Figure 6.8.   Relationship between hyphal length in soil and
percentage of water-stable aggregates greater than 2 mm.

Source: Tisdall and Oades 1979.   Reprinted by permission.

External Hyphae as a Source of Inoculum

Spores, infected root pieces (primarily those containing vesicles), and hyphae may serve as sources of inocula
of VAM fungi. Often, studies on the spread of infection
have failed to distinguish among these infective units.
For example, Powell (1979) found that VAM fungi spread into
nonmycorrhizal seedlings at a rate of 0.6 to 1.5 m/yr;
however, root growth was not assessed independently of
hyphal growth.

Warner and Mosse (1980, 1983) and Warner (1984) conducted a series of experiments using a fabric membrane
system that restricted growth of infected roots but allowed
growth of external hyphae into the surrounding soil. From
these studies they concluded that hyphae alone could ini-
tiate infections, and that the spread of infection from
roots by hyphae occurred over relatively short distances
(10 to 20 mm). Furthermore, hyphae detached from roots lost
viability rapidly, though colonization of organic fragments
extended the infectivity of hyphae. Hyphae in the soil may
be an important source of inoculum in native ecosystems
where large hyphal mats develop (Birch 1986).

METHODS OF QUANTIFICATION

There is no completely satisfactory method for quantifying external hyphae of VAM fungi in soil. Three key
problems have yet to be overcome: (1) reliable method to
distinguish VAM fungi from the myriad of other fungi in the
soil does not exist, (2) assessment of the viability and
activity of hyphae is problematic, and (3) meaningful quantification is very time consuming.

Several indirect measurements of hyphal development
have been attempted. Graham, Linderman, and Menge (1982)
estimated relative hyphal development by weighing plant
roots with adhering soil. They reasoned that the weight of
the root ball would be proportional to the amount of hyphae
that bound the unit together. They obtained an excellent
correlation between their estimate of hyphal development
and plant growth response (Figure 6.5). With this method,
however, it is not possible to compare hyphal development
among different soils or plant species, since many factors
affect the quantity of soil in the root ball. Nicolson
(1959) and Sanders et al. (1977) attempted to weigh hyphae
sieved from soil, but this method suffers from contamina-

tion of hyphae by soil debris, since it is not practical to separate hyphae completely from soil particles. Chitin determinations also have been used to estimate hyphal biomass (Pacovsky and Bethlenfalvay 1982; Ahmadsad 1984). This method has proven very useful under highly controlled experimental conditions. Its utility in native soil, however, is limited, because chitin is ubiquitous in nature, where it is found in the walls of many fungi and in the exoskeletons of insects.

Attempts have been made to estimate hyphal development directly by extracting hyphae from soil and quantifying lengths by a modified gridline-intersect method (Abbott et al. 1984; Sylvia 1986). The major problem with this method is that hyphae of nonmycorrhizal fungi often are not distinguishable from those of VAM fungi. One approach has been to subtract the length of hyphae in control treatments from the length of hyphae in VAM fungi-inoculated treatments; however, this assumes that no interactions occur among the hyphae in the soil.

A common shortcoming of the above methods is that they fail to distinguish the active or functioning portion of the hyphal mass from inactive hyphae. Recently, stains that differentiate actively metabolizing cells from inactive cells have been used to estimate the "viability" of VAM fungal hyphae. Sylvia (1988) found that a solution containing iodonitrotetrazolium (INT) and NADH could be used to locate active hyphae of two Glomus spp.; INT is reduced by the electron-transport system of living cells and results in formation of a red color. Schubert et al. (1987) used fluorescein diacetate (FDA) in their studies to identify functioning hyphae; FDA is hydrolyzed within living cells, releasing fluorescein, which can be detected with a fluorescence microscope. Kough and Gianinazzi-Pearson (1986) used specific stains to detect succinate dehydrogenase, alkaline phosphatase, and lipase activity in mycorrhizal onions. These methods all demonstrate that only a portion of the total fungal mass is physiologically active.

Further development of methods is needed to improve our understanding of the ecophysiology of VAM hyphal systems. Use of polyclonal (Aldwell and Hall 1986; Kough and Linderman 1986) and monoclonal (Morton et al. 1987) antibodies offer promise for monitoring specific fungi in native soil, while computerized image-analysis systems will speed measurement and analysis of data. The use of these techniques should allow for significant advances in our knowledge of the ecology of external hyphae in the near future.

REFERENCES

Abbott, L. K., and A. D. Robson. 1985a. The effect of soil pH on the formation of VA mycorrhizas by two species of Glomus. Australian Journal of Soil Research 23:253-261.

Abbott, L. K., and A. D. Robson. 1985b. Formation of external hyphae in soil by four species of vesicular-arbuscular mycorrhizal fungi. New Phytologist 99:245-255.

Abbott, L. K., A. D. Robson, and G. De Boer. 1984. The effect of phosphorus on the formation of hyphae in soil by the vesicular-arbuscular mycorrhizal fungus, Glomus fasciculatum. New Phytologist 97:437-446.

Ahmadsad, I. 1984. A method for the quantitative determination of the extraradical mycelium of vesicular-arbuscular mycorrhiza. Angewandte Botanik 58:359-364.

Aldwell, F. E. B., and I. R. Hall. 1986. Monitoring spread of Glomus mosseae through soil infested with Acaulospora laevis using serological and morphological techniques. Transactions of the British Mycological Society 87:131-134.

Allen, E. B., and M. F. Allen. 1986. Water relations of xeric grasses in the field: Interactions of mycorrhizas and competition. New Phytologist 104:559-571.

Alexander, C., I. J. Alexander, and G. Hadley. 1984. Phosphate uptake by Goodyera repens in relation to mycorrhizal infection. New Phytologist 97:401-411.

Beever, R. E., and D. J. W. Burns. 1980. Phosphorus uptake, storage and utilization by fungi. In Advances in Botanical Research, ed. H. W. Woolhouse. Vol. 8. 127-219. New York: Academic Press.

Bethlenfalvay, G. J., and R. N. Ames. 1987. Comparison of two methods for quantifying extraradical mycelium of vesicular-arbuscular mycorrhizal fungi. Soil Science Society of America Journal 51:834-837.

Bethlenfalvay, G. J., M. S. Brown, and R. S. Pacovsky. 1982. Relationships between host and endophyte development in mycorrhizal soybeans. New Phytologist 90:537-543.

Bethlenfalvay, G. J., and R. S. Pacovsky. 1983. Light effects in mycorrhizal soybeans. Plant Physiology 73:969-972.

Birch, C. P. D. 1986. Development of VA mycorrhizal infection in seedlings in semi-natural grassland turf. In

Physiological and genetical aspects of mycorrhizae, ed. V. Gianinazzi-Pearson, and S. Gianinazzi. 233-237. Proceedings of the first European Symposium on Mycorrhizae. Paris: INRA.

Bonfante-Fasolo, P. 1984. Anatomy and morphology of VA mycorrhizae. In *VA Mycorrhiza*, ed. C. L. Powell and D. J. Bagyaraj. 5-33. Boca Raton, FL: CRC Press.

Bowen, G. D. 1987. The biology and physiology of infection and its development. In *Ecophysiology of VA mycorrhizal plants*, ed. G. Safir. 27-57. Boca Raton, FL: CRC Press.

Butler, E. J. 1939. The occurrences and systematic position of the vesicular-arbuscular type of mycorrhizal fungi. *Transactions of the British Mycological Society* 22:274-301.

Chiariello, N., J. C. Hickman, and H. A. Mooney. 1982. Endomycorrhizal role for interspecific transfer of phosphorus in a community of annual plants. *Science* 217:941-943.

Cooper, K. M. 1984. Physiology of VA mycorrhizal associations. In *VA mycorrhiza*, ed. C. L. Powell, and D. J. Bagyaraj. 155-186. Boca Raton, FL: CRC Press.

Cooper, K. M., and P. B. Tinker. 1981. Translocation and transfer of nutrients in vesicular-arbuscular mycorrhizas. IV. Effect of environmental variables on movement of phosphorus. *New Phytologist* 88:327-339.

Diem, H. G., M. Gueye, and Y. Dommergues. 1986. The development of extraradical hyphae in relation to the response of cowpea to VA mycorrhizal infection. 227-232. In *Physiological and genetical aspects of mycorrhizae*, ed. V. Gianinazzi-Pearson and S. Gianinazzi. 227-232. Proceedings of the first European Symposium on Mycorrhizae. Paris: INRA.

Douds, D. D., and W. R. Chaney. 1982. Correlation of fungal morphology and development of host growth in a green ash mycorrhizae. *New Phytologist* 92:519-526.

Fitter, A. H. 1985. Functioning of vesicular-arbuscular mycorrhizas under field conditions. *New Phytologist* 99:257-265.

Fitter, A. H. 1986. Effect of benomyl on leaf phosphorus concentration in alpine grasslands: A test of mycorrhizal benefit. *New Phytologist* 103:767-776.

Francis, R., R. D. Finlay, and D. J. Read. 1986. Vesicular-arbuscular mycorrhiza in natural vegetation systems. IV. Transfer of nutrients in inter- and intra-specific combinations of host plants. *New Phytologist* 102:103-111.

Francis, R., and D. J. Read. 1984. Direct transfer of carbon between plants connected by vesicular-arbuscular mycorrhizal mycelium. Nature 307:53-56.

Gianinazzi-Pearson V., and H. G. Diem. 1982. Endomycorrhizae in the tropics. In Microbiology of tropical soil and plant productivity, ed. Y. R. Dommergues and H. G. Diem. 328-351. The Hague: Martinus Nijhoff.

Graham, J. H. 1987. Non-nutritional benefits of VAM fungi--do they exist? In Mycorrhizae in the next decade: Practical applications and research priorities, ed. D. M. Sylvia, L. L. Hung, and J. H. Graham. 237-239. Proceedings of the seventh North American conference on Mycorrhizae. Gainesville, FL: Institute of Food and Agricultural Sciences.

Graham, J. H., R. G. Linderman, and J. A. Menge. 1982. Development of external hyphae by different isolates of mycorrhizal Glomus spp. in relation to root colonization and growth of Troyer citrange. New Phytologist 91:183-189.

Graham, J. H., L. W. Timmer, and D. Fardelmann. 1986. Toxicity of fungicidal copper in soil to citrus seedlings and vesicular-arbuscular mycorrhizal fungi. Phytopathology 76:66-70.

Graham, J. H., and J. P. Syvertsen. 1984. Influence of vesicular-arbuscular mycorrhiza on the hydraulic conductivity of roots of two citrus rootstocks. New Phytologist 97:277-284.

Grime, J. P., J. M. L. Mackey, S. H. Hillier, and D. J. Read. 1987. Mechanisms of floristic diversity: A key role for mycorrhizae. In Mycorrhizae in the next decade: Practical applications and research priorities, ed. D. M. Sylvia, L. L. Hung, and J. H. Graham. 151. Proceedings of the seventh North American conference on mycorrhizae. Gainesville, FL: Institute of Food and Agricultural Sciences.

Hardie, K., and L. Leyton. 1981. The influence of vesicular-arbuscular mycorrhiza on growth and water relations of red clover. I. In phosphate deficient soil. New Phytologist 89:599-608.

Harley, J. L., and E. L. Harley. 1987. A check-list of mycorrhiza in the British flora. New Phytologist Supplement 105:1-102.

Harley, J. L., and S. E. Smith. 1983. Mycorrhizal symbiosis. New York: Academic Press.

Hayman, D. S. 1978. Endomycorrhizae. In Interactions between non-pathogenic soil microorganisms and plants, ed. Y.R. Dommergues and S.V. Krupa. 401-442. Oxford, England: Elsevier.

Heap, A. J., and E. I. Newman. 1980. The influence of vesicular-arbuscular mycorrhizas on phosphorus transfer between plants. New Phytologist 85:173-179.

Heckman, J. R., and J. S. Angle. 1987. Variation between soybean cultivars in vesicular-arbuscular mycorrhiza fungi colonization. Agronomy Journal 79:428-430.

Hirrel, M. C., and J. W. Gerdemann. 1979. Enhanced carbon transfer between onions infected with a vesicular-arbuscular mycorrhizal fungus. New Phytologist 83:731-738.

Jeffries, P. 1987. Use of mycorrhizae in agriculture. CRC Critical Reviews in Biotechnology 5:319-357.

Kjøller, A., and S. Struwe. 1982. Microfungi in ecosystems: Fungal occurrence and activity in litter and soil. Oikos 39:391-422.

Koske, R. E., J. C. Sutton, and B. R. Sheppard. 1975. Ecology of Endogone in Lake Huron sand dunes. Canadian Journal of Botany 53:87-93.

Kough, J. L., and V. Gianinazzi-Pearson. 1986. Physiological aspects of VA mycorrhizal hyphae in root tissue and soil. In physiological and genetical aspects of mycorrhizae, ed. V. Gianinazzi-Pearson and S. Gianinazzi. 223-226. Proceedings of the first European Symposium on Mycorrhizae. Paris: Institut National de Recherches Agronomiques.

Kough, J. L., N. Majajczuk, and R. G. Linderman. 1986. Use of the indirect immunofluorescent technique to study the vesicular-arbuscular fungus Glomus etunicatum and other Glomus species. New Phytologist 94:57-62.

Kough, J. L., and R. G. Linderman. 1986. Monitoring extra-matrical hyphae of a vesicular-arbuscular mycorrhizal fungus with an immunofluorescence assay and the soil aggregation technique. Soil Biology and Biochemistry 18:307-313.

Kucey, R. M. N., and E. A. Paul. 1982. Biomass of mycorrhizal fungi associated with bean roots. Soil Biology and Biochemistry 14:413-414.

Lynch, J. M., and E. Bragg. 1985. Microorganisms and soil aggregate stability. Advances in Soil Science 2:133-171.

Morton, J. B., S. F. Wright, and J. E. Sworobuk. 1987. Sensitivity of monoclonal antibodies to detect propagules of Glomus occultum. In Mycorrhizae in the next decade: Practical applications and research priorities, ed. D. M. Sylvia, L. L. Hung, and J. H. Graham. 317. Proceedings of the seventh North American conference on mycorrhizae. Gainesville, FL: Institute of Food and Agricultural Sciences.

Moser, M., and K. Haselwandter. 1983. Ecophysiology of mycorrhizal symbionts. In Physiological plant ecology III, ed. O. L. Lange et al. 391-422. New York: Springer-Verlag.

Mosse, B. 1959. Observations on the extra-matrical mycelium of a vesicular-arbuscular endophyte. Transactions of the British Mycological Society 42:439-448.

Mosse, B., D. P. Stribley, and F. LeTacon. 1981. Ecology of mycorrhizae and mycorrhizal fungi. In Advances in Microbial Ecology, vol. 5, 137-210, ed. M. Alexander New York: Plenum Publishing Co.

Nelson, C. E. 1987. The water relations of vesicular-arbusular mycorrhizal systems. In Ecophysiology of VA Mycorrhizal Plants, ed. G. R. Safir. 71-91. Boca Raton, FL: CRC Press.

Newman, E. I., and K. Ritz. 1986. Evidence on the pathways of phosphorus transfer between vesicular-arbuscular mycorrhizal plants. New Phytologist 104:77-87.

Nicolson, T. H. 1959. Mycorrhizae in the Gramineae. I. Vesicular-arbuscular endophytes with special reference to the external phase. Transactions of the British Mycological Society 42:421-438.

Nicolson, T. H., and C. Johnston. 1979. Mycorrhizae in the Gramineae. III. Glomus fasciculatum as the endophyte of pioneer grasses in a maritime sand dune. Transactions of the British Mycological Society 72:261-268.

Owusu-Bennoah, E., and A. Wild. 1979. Autoradiography of the depletion zone of phosphate around onion roots in the presence of vesicular-arbuscular mycorrhiza. New Phytologist 82:133-140.

Pacovsky, R. S., and G. J. Bethlenfalvay. 1982. Measurement of the extraradical mycelium of a vesicular-arbuscular mycorrhizal fungus in soil by chitin determination. Plant and Soil 68:143-147.

Powell, C. L. 1976. Development of mycorrhizal infections from Endogone spores and infected root segments. Transactions of the British Mycological Society 66:439-445.

Powell, C. L. 1979. Spread of mycorrhizal fungi through soil. New Zealand Journal of Agricultural Research 22:335-341.

Read, D. J. 1984. The structure and function of vegetative mycelium of mycorrhizal roots. In Ecology and Physiology of the Fungal Mycelium, ed. D. H. Jennings and A. D. M. Rayner. 215-240. Cambridge, England: Cambridge University Press.

Rhodes, L. H., and J. W. Gerdemann. 1975. Phosphate uptake zones of mycorrhizal and non-mycorrhizal onions. New Phytologist 75:555-561.

St. John, T. V., D. C. Coleman, and C. P. P. Reid. 1983. Association of vesicular arbuscular mycorrhizal hyphae with soil organic particles. Ecology 64:957-959.

Same, B. I., A. D. Robson, and L. K. Abbott. 1983. Phosphorus, soluble carbohydrates and endomycorrhizal infection. Soil Biology and Biochemistry 15:593-597.

Sanders, F. E., and N. A. Sheikh. 1983. The development of vesicular-arbuscular mycorrhizal infection in plant root systems. Plant and Soil 71:223-246.

Sanders, F. E., and P. B. Tinker. 1973. Phosphate flow into mycorrhizal roots. Pesticide Science 4:385-395.

Sanders, F. E., P. B. Tinker, R. L. B. Black, and S. M. Palmerley. 1977. The development of endomycorrhizal root systems: I. Spread of infection and growth-promoting effects with four species of vesicular-arbuscular endophyte. New Phytologist 78:257-268.

Scheltema, M. A., L. K. Abbott, A. D. Robson, and G. De'ath. 1987. The spread of mycorrhizal infection by Gigaspora calospora from a localized inoculum. New Phytologist 106:727-734.

Schubert, A., C. Marzachi, M. Mazzitelli, M. C. Cravero, and P. Bonfante-Fasolo. 1987. Development of total and viable extraradical mycelium in the vesicular-arbuscular mycorrhizal fungus Glomus clarum Nicol. & Schenck. New Phytologist 107:183-190.

Schwab, S. M., J. A. Menge, and R. T. Leonard. 1983. Comparison of stages of vesicular-arbuscular mycorrhiza formation in Sudangrass grown at two levels of phosphorus nutrition. American Journal of Botany 70:1225-1232.

Skinner, M. F., and G. D. Bowen. 1974. The penetration of soil by mycelial strands of ectomycorrhizal fungi. Soil Biology and Biochemistry 6:57-61.

Smith, S. E., and N. A. Walker. 1981. A quantitative study of mycorrhizal infection in Trifolium: Separate determination of rates of infection and of mycelial growth. New Phytologist 89:225-240.

166

Sutton, J. C. 1973. Development of vesicular-arbuscular mycorrhizae in crop plants. Canadian Journal of Botany 51:2487-2493.

Sutton, J. C., and B. R. Sheppard. 1976. Aggregation of sand-dune soil by endomycorrhizal fungi. Canadian Journal of Botany 54:326-333.

Sylvia, D. M. 1986. Spatial and temporal distribution of vesicular-arbuscular mycorrhizal fungi associated with Uniola paniculata in Florida foredunes. Mycologia 78:728-734.

Sylvia, D. M. 1988. Activity of external hyphae of vesicular-arbuscular mycorrhizal fungi. Soil Biology and Biochemistry 20:39-43.

Sylvia, D. M., and J. N. Burks. 1988. Selection of vesicular-arbuscular mycorrhizal fungi for inoculation of Uniola paniculata L. Mycologia 80:565-568.

Tinker, P. B., and A. Gildon. 1983. Mycorrhizal fungi and ion uptake. In Metals and micronutrients, uptake and utilization by plants, ed. D. A. Robb and W. S. Pierpoint. 21-32. New York: Academic Press.

Tisdall, J. M., and J. M. Oades. 1979. Stabilization of soil aggregates by the root segments of ryegrass. Australian Journal of Soil Research 17:429-441.

Trappe, J. M. 1987. Phylogenetic and ecological aspects of mycotrophy in the angiosperms from an evolutionary standpoint. In Ecophysiology of VA mycorrhizal plants, ed. G.R. Safir. 5-25. Boca Raton, FL: CRC Press.

van Kessel, C., P. W. Singleton, and H. J. Hoben. 1985. Enhanced nitrogen-transfer from a soybean to maize by vesicular arbuscular mycorrhizal (VAM) fungi. Plant Physiology 79:562-563.

Viets, F. G. 1972. Water deficits and nutrient availability. In Water deficits and plant growth, vol. 3. ed. T. T. Kozlowski, 217-239. New York: Academic Press.

Warner, A. 1984. Colonization of organic matter by vesicular-arbuscular mycorrhizal fungi. Transactions of the British Mycological Society 82:352-354.

Warner, A., and B. Mosse. 1980. Independent spread of vesicular-arbuscular mycorrhizal fungi in soil. Transactions of the British Mycological Society 74:407-410.

Warner, A., and B. Mosse. 1983. Spread of vesicular-arbuscular mycorrhizal fungi between separate root systems. Transactions of the British Mycological Society 80:353-354.

Warnock, A. J., A. H. Fitter, and M. B. Usher. 1982. The influence of a springtail *Folsomia candida* (Insecta, Collembola) on the mycorrhizal association of leek, *Allium porrum* and the vesicular-arbuscular mycorrhizal endophyte *Glomus fasciculatum*. New Phytologist 90:285-292.

Whittingham, J., and D. J. Read. 1982. Vesicular-arbuscular mycorrhizae in natural vegetation systems. III. Nutrient transfer between plants with mycorrhizal interconnections. New Phytologist 90:277-284.

Wilson, J. M. 1984. Comparative development of infection by three vesicular-arbuscular mycorrhizal fungi. New Phytologist 97:413-426.

R. J. Snider
R. Snider
A. J. M. Smucker

7 Collembolan Populations and Root Dynamics in Michigan Agroecosystems

Collembola are known to be significant regulators of the decomposition of soil organic matter. They have also been shown to reduce plant growth in agroecosystems through mycorrhizal grazing (Warnock, Fitter, and Usher 1982) and direct root consumption (Brown 1985). These associations with root systems have remained obscure, however, primarily due to the paucity of methods available to quantitatively describe their activities.

Collembolan populations have often been shown to fluctuate in a bimodal seasonal pattern: declines are most pronounced in mid-winter and mid-summer, while peaks tend to occur in spring and fall (Wallwork 1970). These data were generally obtained by conventional methods; that is, heat extraction of soil cores taken from depths of 30 cm or less. Consequently, deep-dwelling populations were excluded. In addition, heat extraction of mesofaunca can be very inefficient. The numbers of animal populations extracted often vary with both species and developmental state (Petersen and Luxton 1982; Van Straalen and Rijninks 1982).

MINIRHIZOTRON AND VIDEO IMAGES OF THE RHIZOSPHERE

Minirhizotron observation of root dynamics by the use of minirhizotron has become an accepted method for quantifying root characteristics (i.e., branching, frequency, rate of elongation, root depth, and death. This nondestructive approach for quantifying root turnover rates and repeated observations of the root and soil interface has greatly contributed to the discipline of rhizodynamics

(Taylor 1987). Microvideo color cameras are used to record rhizosphere activities at the soil-minirhizotron interface. Modifications of commercially available cameras have been used to quantify root branching and nodulation (Ferguson and Smucker 1989). In addition, large populations of Collembola can be observed in each window, 2.16 cm², at intervals of 1.2 cm to soil depths of 150 cm. Repeated measurements at each depth can be taken at frequent time intervals throughout the year. This chapter summarizes data collected during the 1985 and 1986 growing seasons at the Michigan State University research farms.

Crops were grown on two soil types for a two-year period. Corn (Zea mays L.) and soybeans (Glycine max L.), planted on a Metea sandy loam (Arenic Hapludalf) and a Capac loam (Aeric Ochraqualf), and sugarbeets (Beta vulgaris L.), planted only on the Metea sandy loam, were evaluated for root activity and Collembola populations during a series of high-yield agronomic experiments.

Soil and its components, adjacent to the minirhizotron were video recorded by a modified video camera system described by Ferguson and Smucker (1989) and Upchurch and Ritchie (1984). Video recordings, at each 1.2 cm interval, were recorded on high quality 3/4 in videotape for 3 s without moving the camera. This approach provides opportunities for reproducible and precise evaluations of root and associated mesofaunal activities (Ferguson and Smucker 1989). Video recordings of roots were taken from minirhizotrons installed at 45 degree angles immediately after planting according to the methods described by Box, Smucker, and Ritchie (1989). Roots appearing in each video image were counted manually and summed by intervals of 10 cm. Root turnover rates, the number of roots per cm² per day, were calculated from the differences in the numbers of roots appearing at the composite of twelve windows from eight minirhizotron tubes for each date and divided by the surface area of this composite. Occasionally, root numbers at a given position or "window" along the tube were significantly greater than the average of the replicates. Therefore, a smoothing algorithm*, which excluded values greater than one standard deviation, was developed and applied to the root data sets. During the manual counting of the roots for these experiments, we concluded that Collembola populations in these video-recorded images should also be quantified.

The number of roots per cm² was determined by counting

---

*Algorithm available from the authors on request.

all roots in each image, regardless of root size, and dividing the total number by 2.16 cm². In this manner, all roots were accessed equally. Vertical distribution of Collembola in the rhizospheres of soybean, sugarbeet, and corn crops were also assessed manually. If the number of animals were small, we counted them by visually scanning the entire video screen. Images with too many Collembola to count during the three-second exposure time were covered with an opaque mask that covered 1/2 to 3/4 of the video screen; the exposed animals were counted, and then the cover was reversed, the video tape replayed, and the remaining animals counted. In these cases, each video image was viewed two and four times, respectively, and the total number of animals per frame was obtained by summation. In this manner, populations of Collembola could be quantified from the video-recorded images of the minirhizotron system (Figure 7.1A).

Depth of focus and sharpness of these tapes were often very good (Figure 7.1B and C). There were marginal images, however, which interfered with the detection of very small individuals (Figure 7.1D-F). In addition, frames advanced at three-second intervals and could not be paused without fine resolution being lost. As a result, only moving Collembola were positively identified. Since motion was a prerequisite for identifying Collembola, data presented here probably underestimate the total collembolan numbers. The slow-moving and often very small Onychiuridae were either not seen at all or were purposely ignored because their true numbers could not be assessed with confidence. Of the two species that could be reliably recognized, _Flsomia candida_ (Willem) was by far the dominant species; rare individuals of _Pseudosinella violenta_ (Folsom)--identified from large individuals in sharp focus)--were included in final counts. Therefore, these data pertain to the distribution of _F. candida_ in three agroecosystems.

The number of images from which average numbers of Collembola were derived was variable, limited by image quality (sharpness) and the maximum depth at which video taping began. It should be reiterated that these tapes were generated for root-growth studies and were not made with animal observations in mind. Thus, while visual records to a vertical depth of approximately 1 m were available for mid-summer, tapes made early in the season generally did not extend below 30 or 40 cm. In addition, not all video images were taken at equal depths for each date, so replication was generally reduced with increasing depth. Counts of Collembola were summed over consecutive

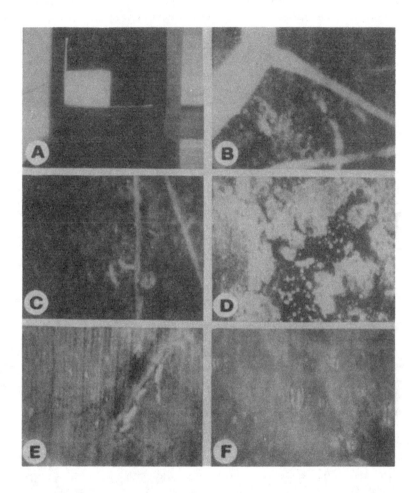

Figure 7.1   Video images from minirhizotron tubes: (A)
video monitor screen with an opaque mask that aided in the
counting of images containing large numbers of Collembola,
(B) image of corn roots on August 20, 1985, at 46 cm depth
in Capac soil, (C) image of soybean roots and Collembola on
August 20, 1985, at 43 cm depth in Metea soil, (D) partially obscured image of corn root on August 20, 1985, at 72 cm
depth in Metea soil, (E) partially obscured image of soybean roots and Collembola on March 12, 1986, at 36 cm depth
in Metea soil, and (F) image obscured by condensation of
water on minirhizotron tube on April 12, 1985, in a Metea
soil.

images at three contiguous depths. Results are thus presented as the mean number of Collembola per 6.48 cm^2 with
means based on a variable number of minirhizotron observations for each date and treatment.

EXTRACTION OF COLLEMBOLA FROM SOIL CORES

Destructive soil samples were taken from Metea and
Capac soils in mid-September 1986. Samples 5 cm in
diameter were taken with a hydraulic Giddings probe
(Giddings Co., Ames, Iowa). Although vertical compression
of the soil was unavoidable, samples were taken to depths
of 120 cm. Cores were cut into 10-cm sections to a depth
of 80 cm, then into two additional 20-cm sections (at
depths of 80 to 100 and 100 to 120 cm). They were heat-
extracted in Tullgren funnels for five days, and only
Collembola, F. candida, were sorted to species and counted.

Collembola Dynamics

Collembolan activities associated with the rhizospheres of three crops were observed at greater depths than
previously reported. Many more **Folsomia candida** were observed, by the minirhizotron technique, in the rhizospheres
of soybeans and sugarbeets than in the rhizospheres of corn
(Figures 7.2, 7.3, and 7.4). Collembola were the most
numerous in the upper portions of the soybean soil profile
by mid-July (Figure 7.4). Additionally, a downward migration of Collembola occurred as the season progressed
(Figures 7.2-7.4). The scarcity of soil animals in the
upper 30 cm later in the season may have resulted from the
progression of roots into greater soil depths (reported
later) or some other factors, e.g., temperature, which were
not measured. August was also the month when conspicuously
large numbers of immature Collembola appeared randomly at
soil depths to at least 90 cm. Activities of these young
animals appeared to be greater at 40- to 90-cm depths, and
they were frequently located in soil areas devoid of roots.

In fall and winter, populations restratified toward
the upper half of the profile (Figure 7.5). Although their
distribution may have been bimodal in winter with a second
peak occurring below 60 cm on January 8, available evidence
indicates a pronounced concentration of animals in the 20-

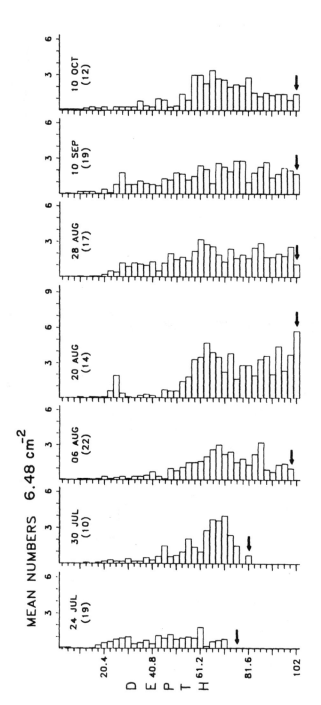

Figure 7.2. Collembolan population profiles in a sugarbeet and Metea soil agroecosystem during the production season of 1986. Values in parentheses are the replications for each date. Arrows indicate the maximum depth of measurement for that date.

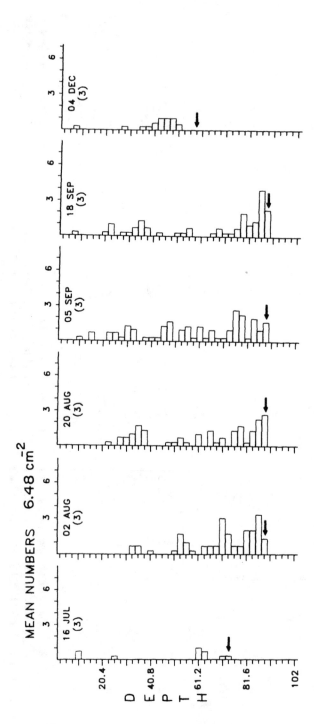

Figure 7.3. Collembolan population profiles in a corn and Metea soil agroecosystem during the production season of 1985. (See Figure 7.2 for indicators.)

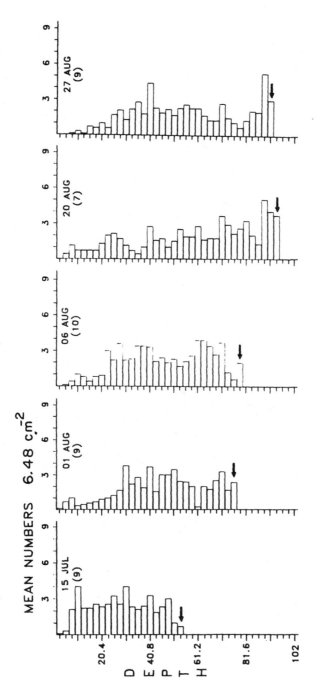

Figure 7.4. Collembolan population profiles in a soybean and Metea soil agroecosystem during the production season of 1985. (See Figure 7.2 for indicators.)

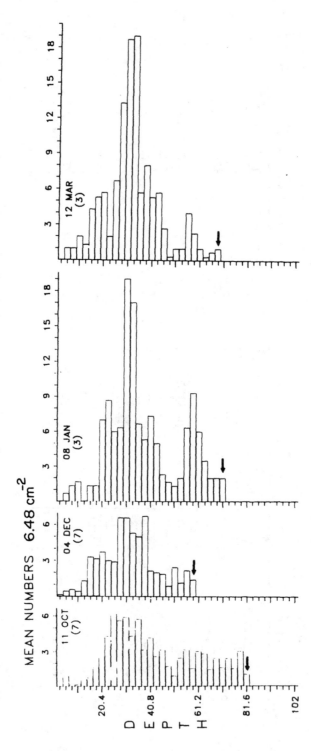

Figure 7.5. Collembolan population profiles in a soybean and Metea soil agroecosystem during the autumn and winter of 1985 and 1986. (See Figure 7.2 for indicators.)

to 40-cm stratum (e.g., January and March 1986, Figure 7.5).

The tendency for Collembola to become deep dwelling in late summer was equally evident in sugarbeet crops where the upper 40 cm of the soil profile was very poorly populated in late July and early August (Figure 7.2). According to videos made on October 10, this vertical stratification pattern appeared more persistent here than under soybean (Figure 7.4).

Numbers of Collembola differed with soil type as well as with crop. In Metea loamy sand, the corn rhizosphere harbored fewer animals than either soybean or sugarbeet (Figure 7.3). In Capac loam, soybean (Figure 7.6) also promoted somewhat larger populations than did corn (Figure 7.7). Interpretation of seasonal vertical distribution becomes tenuous, however, as animal numbers become lower. The great depth to which Collembola migrate, particularly at the height of the growing season was clear in all treatments.

The number of F. candida extracted from soil cores was greater under soybean than under corn in both soil types (Table 7.1). In samples from Metea loamy sand plots where the water table was high, none were found below 60 cm. In Capac loam, F. candida ranged to depths of 80 to 100 cm.

TABLE 7.1. Total number of Folsomia candida in each 203 cm³ of soil as measured by the heat extraction method, September 1986.

Depth (cm)	Capac loam		Metea loamy sand	
	Soybean	Corn	Soybean	Corn
0-10	63	4	82	1
10-20	10	6	53	13
20-30	5	-	11	--
30-40	2	1	2	3
40-50	2	2	--	2
50-60	15	2	--	2
60-70	3	-	--	--
70-80	1	4	--	--
80-100	--	2	--	--

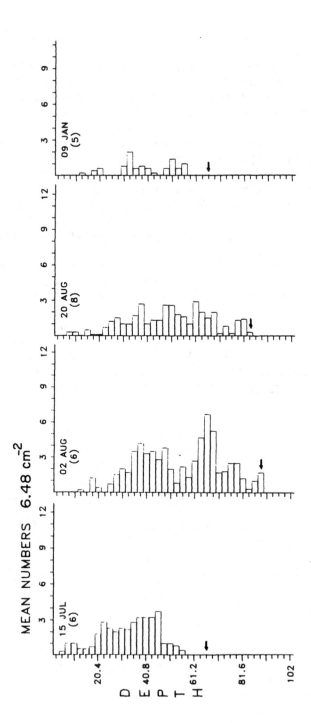

Figure 7.6 Collembolan population profiles in a soybean and Capac soil agroecosystem during the production season of 1985. (See Figure 7.2 for indicators.)

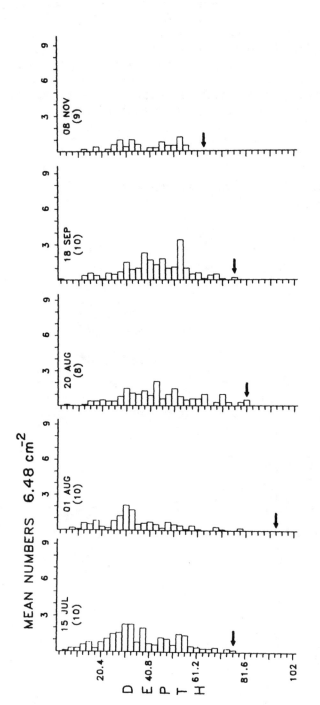

Figure 7.7 Collembolan population profiles in a corn and Capac soil agroecosystem during the production season of 1985. (See Figure 7.2 for indicators.)

180

Since the soil cores were severely compacted, extraction efficiency was presumably very poor, particularly for the subsamples deeper in the profile. Relative numbers of animals per depth increment are therefore of little relevance. The data indicate, however, that the species is indeed distributed throughout the profile to depths of approximately 1 m, qualitatively verifying the Collembola distributions observed by the minirhizotrons.

Root Dynamics

Estimation of root system activities by the minirhizotron method at four dates after planting suggest large

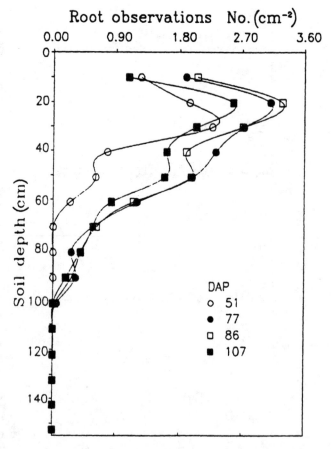

Figure 7.8. Root development profiles of soybeans in a Metea soil, 1986.

differences among the root profiles of different cultivars (Figures 7.8-7.10). Roots of corn and soybeans were concentrated in the upper 40 cm of the soil, while sugarbeet roots were distributed throughout the upper 80 cm. The horizon interface between the A and B soil horizons, at 45 cm in the Metea soil series, dramatically reduced the number of roots of all cultivars in this portion of the soil profile. It is worth noting, however, that this interface appeared to have little influence on the root growth of sugarbeets 105 days after planting (DAP) (Figure 7.10). In contrast, the interface between horizons B and C, at 70 cm, influenced the accumulations of corn roots at 111 DAP and sugarbeet roots at 71, 84, and 105 DAP (Figures 6.9 and 7.10). Maximum root numbers in the 0- to 20-cm portions of the soil profiles occurred 86 and 90 DAP for

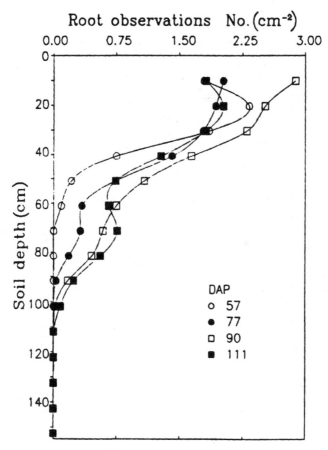

Figure 7.9. Root development profiles of corn in a Metea soil, 1986.

soybeans and corn. Maximum accumulations of sugarbeet roots occurred at 50 cm at 105 DAP. Significant losses of soybean roots occurred, to depths of 60 cm, during the period from 86 to 107 DAP (Figure 7.8).

Root activities, defined as the change in the number of roots divided by the time between measurements, portray the dynamics of root development at 10-cm intervals throughout the soil profile (Figures 7.11-7.13). Numbers greater than zero indicate root production rates were greater than death rates; and numbers less than zero indicate root death rates were greaterthan production rates. The greatest growth of roots occurred during the periods from 71 to 77, 77 to 83, and 84 to 90 DAP for the soybeans, corn, and sugarbeets, respectively. These maximum growth rates were

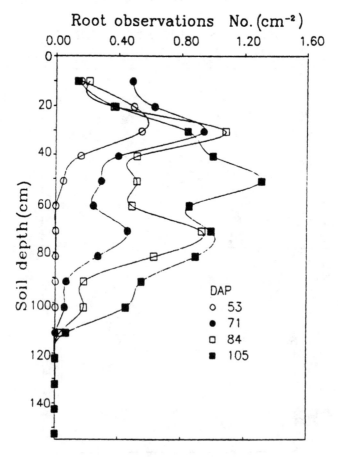

Figure 7.10. Root development profiles for sugarbeets in a Metea soil, 1986.

species dependent and varied with respect to soil depth. The greatest growth of corn roots occurred in the upper 30 cm. The greatest growth of soybean roots occurred throughout the soil profile to depths of 80 or 90 cm (Figure 7.11). The most activity of sugarbeet roots occurred at depths of 35 to 60 cm for the period from 84 to 90 DAP and at depths of 60 to 75 cm, 71 to 76 DAP. The greatest death rates of roots differed with time, cultivar, and soil depth. Generally, root death rates were greatest in areas of the soil that had accumulated the most roots. Soybeans exhibited the highest death rates, up to 1250 m^{-2} d^{-1} for the period from 51 to 71 DAP in the upper 30 cm of the soil

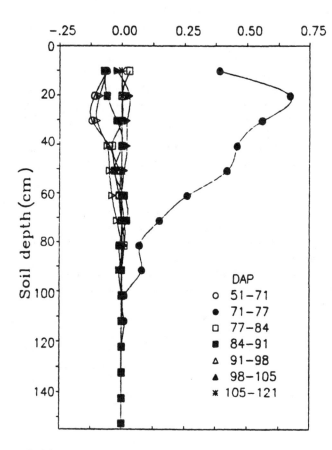

Figure 7.11. Root activity profiles for soybeans in a Metea soil, 1986.

profile (Figure 7.11). Maximum corn root losses were approximately 50 percent of those of soybean, up to 750 m^{-2} d^{-1}, in the upper 30 cm of the soil profile 90 to 97 DAP. Maximum sugarbeet root losses were approximately 300 m^{-2} d^{-1} at the 30- and 50- cm depths for the periods from 84 to 90 and 90 to 97 DAP, respectively (Figure 7.13). Concurrent root death in the upper portions of the soil profile and root growth at greater soil depths during the same time periods suggest contrasting root responses to increasing soil moisture gradients between the drier surface soil to the moist soil deeper in the profile (Figures 7.12 and 7.13).

Roots and Collembola. Minirhizotron evidence suggests a positive but low correlation between root numbers and Collembola. Video recorded images showed that Collembola were absent when rootswere absent, while those with numerous roots generally showed significantly greater collembolan activity. Regression correlations between the number of Collembola and the number of soybean roots per frame were very low, however (Table 7.2). The negative correlations between root numbers and soil depth, which contrast with the positive correlations between collembolan numbers and soil depth, support a lack of a good correlation between the numbers of Collembola and plant roots. Additional reasons for these low correlations between Collembola and root numbers include the short measurement time captured by each video image and the small depths of field

TABLE 7.2. Regression correlations between the numbers of Collembola and root densities at 10-cm intervals to a soil depth of 100 cm for a period of 35 d in a soybean field (P = 0 . 0 5).

Date		Collembola vs. roots	Roots vs. soil depth	Collembola vs. soil depth
	(n)	-----------------	r values	-----------------
July 16	(16)	0.360	-0.777	0.383
August 2	(28)	0.289	-0.770	0.301
August 6	(67)	0.055	-0.163	0.119
August 20	(21)	0.240	-0.682	0.335

that could exclude other Collembola in the vicinity of the
rhizosphere. The presence of Collembola in frames without
roots begs similar but obverse arguments. The presence and
density of the soil matrix adjacent to the minirhizotron is
also not taken into account as a potential factor in obscur-
ing the distribution of these mobile but nonburrowing
animals. Finally, other determinants of collembolan activity, notably the density and distribution of the microflora, are ignored when simple numerical coincidences of
roots and Collembola are analyzed.

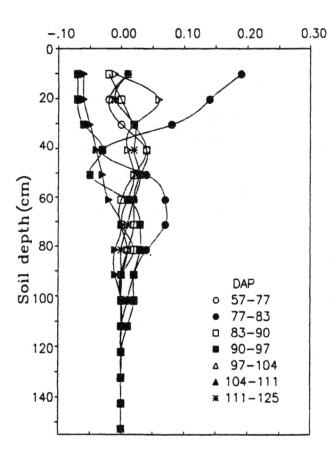

Figure 7.12.   Root activity profiles of corn in a Metea
soil, 1986.

Average numbers of Collembola at each soil depth were much lower for corn than for soybeans (Table 7.3). Interestingly, the lower populations of animals observed in soil planted to corn increased with greater soil depths. Total collembolan numbers for the measured soil profile were 511 percent greater in the soybean agroecosystem than in corn. Combining these total numbers of animals and roots, we can calculate the number of roots observed for each Collembola for the soil depths from 0 to 80 cm. Assuming that plant roots directly or indirectly contribute to the food supply

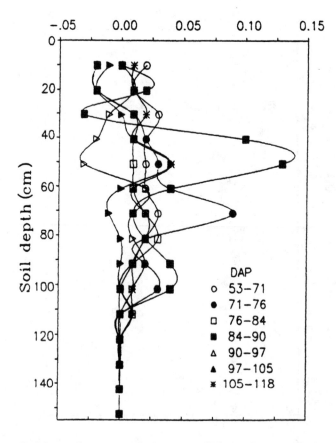

Figure 7.13. Root activity profiles for sugarbeets in a Metea soil, 1986.

of Collembola, it is interesting to note that each animal required 22.2 corn roots and 3.6 soybean roots, or 620 percent more roots are required to sustain a Collembolan in corn than in soybean agroecosystems. Although the mechanisms of Collembola and plant root associations are unknown, these contrasting relationships between the animal and plant root systems are noteworthy and support the direct interactions between soil microarthropods and higher plants of agroecosystems (Finlay 1985).

DISCUSSION

Despite the postulated inefficiency of heat extraction for compacted soil cores, our data indicate that Collembola frequent soils at greater depths than the 30 cm previously reported (Wallwork 1970). The mere presence of individuals of F. candida to depths of 80 to 100 cm, combined with the positive correlation with data by heat extraction, qualitatively support our assumption that Collembola populations in the rhizosphere when observed by the minirhizotron meth-

TABLE 7.3. Average collembolan and root distribution in a Metea loam soil containing soybeans (Glycine max, L.) and corn (Zea mays, L.) during the vegetative growth stages (n - 3-8).

Soil depth	Collembola		Roots	
	Soybeans	Corn	Soybeans	Corn
cm	------------No. per 10 cm^2-----------------			
0-10	2.69±0.92	0.22±0.10	26.7±3.0	52.0±6.2
10-20	6.56±2.50	0.17±0.12	33.5±2.6	37.7±2.7
20-30	9.61±1.50	1.03±0.44	36.1±5.1	42.8±3.5
30-40	7.56±1.50	1.67±0.55	27.0±5.5	47.2±2.8
40-50	6.31±0.75	0.80±5.00	30.0±2.3	30.1±1.5
50-60	6.78±1.00	1.82±0.47	20.7±9.3	20.8±0.7
60-70	9.38±2.00	1.93±0.61	14.0±2.1	7.0±0.9
70-80	7.21±1.75	3.30±0.70	12.7±1.5	6.2±1.8
TOTAL	56.10	10.97	200.70	243.80

od is valid and that observed collembolan distributions were not induced by the installation or presence of mini-rhizotron tubes in the soil. Furthermore, our data indicate that the migration of collembolan numbers throughout the surface meter of the soil profile may be easily misinterpreted when compared to traditional sampling methods. Soil cores, taken to depths of 30 cm or less, appear to yield only a small fraction of any given population due to the shallow depth and possible compaction of the soil sample.

Previously published seasonal fluctuations in numbers of Collembola, which often include winter and summer declines, have generally been attributed to cooler temperatures. In light of present data, we suggest that animal migrations from and into the upper 30 cm, which has been sampled in the past, offer an additional explanation for these fluctuations.

Although based on circumstantial evidence not yet quantifiable, present data indicate that a direct relationship exists between collembolan activity and root development or decomposition, with positive correlation between root activity and the stratification of animal populations.

In conclusion, although the minirhizotron was developed for documenting root growth and turnover, it has contributed to our understanding of collembolan activities. It must, however, be modified (i.e., longer exosure times and ultraviolet light*) to better serve Collembola research. We also offer a summary critique of the new technique and comment on certain aspects of data interpretation currently imposed by its use:

- Tubes may become obscured by the condensation of water and microfloral growth so that even large animals in motion appear only as uncertain shadows. Videos made during the fall, winter, and spring are particularly prone to condensation inside the tube making them temporarily unusable.
- Care must be taken that the camera is precisely focused.
- Length of exposure at each depth should be longer than three seconds to allow Collembola to move and be more easily recognized. Although we suggest

---

*Currently available from Bartz Technology, Santa Barbara, CA.

an exposure of approximately ten seconds, optimal exposure time will be a compromise between the accuracy of counts and the total time available for taping a desired number of minirhizotron sequences.

- Time for viewing tapes and counting Collembola could be much reduced if the videos were subjected to computerized image analysis.
- Both exposure time and computerized enumeration of animals should be customized for the particular species present in a given soil. Unlike <u>F. candida</u>, which are in almost perpetual motion, tiny, slow Onychiuridae (e.g., <u>Tullbergia</u> spp.) may require longer exposure and careful validation by image analysis.
- For Collembola (although not for roots), visual records should reach equal depths (to 1 m or more) throughout the year, yielding complete profiles of vertical distribution.
- Minirhizotron tubes were found to vary with respect to the number of surrounding animals. Many had large populations associated with them all year long; others were consistently poor in fauna.

Minirhizotron records provide a unique opportunity for researchers to view in situ the activities of soil animals. At the current state of development, however, resulting data consist essentially of counts of animals present in small volumes of soil. They do not yield reliable estimates of population density, although that may become possible through technical adjustments and data validation by other sampling methods.

The preliminary data presented here illustrate, however, the potential of the technique for quantifying collembolan distribution in relation to depth, root development, soil type, and climate; investigating population parameters such as size frequencies, recruitment, growth, and mortality; evaluating, in relative terms, differences between treatments, soils, or entire agroecosystems using Collembola as an "indicator standard." These data can lead to a number of fundamental questions on the true functional position of Collembola within the root and soil system and the web of root/soil/microflora interactions.

REFERENCES

Box, J. E., A. J. M. Smucker, and J. T. Ritchie. 1989. Minirhizotron installation techniques for investigating root responses to drought and oxygen stresses. Soil Science Society of America Journal 53:115-118.

Brown, R. A. 1985. Effects of some root-grazing arthropods on the growth of sugarbeets. In Ecological interactions in soil, ed. A. H. Fitter, D. Atkinson, D. J. Read, and M. B. Usher, 285-295. Oxford, England: Blackwell Science Publishers.

Coleman, D. C., and A. Sasson. 1980. Decomposer subsystem. In Grasslands, systems analysis and man, ed. A. I. Breymeyer and G. M. van Dyne 610-648. I.B.P. 19. Cambridge, England: Cambridge University Press.

Ferguson, J. C., and A. J. M. Smucker. 1989. Modifications of the minirhizotron video camera system for measuring spatial and temporal root dynamics. Soil Science Society of America Journal 53:1601-1605.

Finlay, R. D. 1985. Interactions between soil microarthropods and endomycorrhizal associations of higher plants. In Ecological interactions in soil, plants microbes and animals, ed. A. H. Fitter, D. Atkinson, D. J. Read, and M. B. Usher. Special Publication British Ecology Society no. 4:319-331. Oxford, England: Blackwell Science Publishers.

Petersen, H., and M. Luxton. 1982. A comparative analysis of soil fauna populations and their role in decomposition processes. Oikos 39:287-376.

Scott, J. A., N. R. French, and J. W. Leetham. 1979. Patterns of consumption in grassland ecology. Ecological Studies 32. Berlin: Springer-Verlag.

Taylor, H. M., ed. 1987. Minirhizotron observation tubes: Methods and applications for measuring rhizosphere dynamics. American Society of Agronomy Special Publication no. 50. Madison, WI: American Society of Agronomy.

Upchurch, D. R., and J. T. Ritchie. 1984. Battery operated color video camera for root observations in minirhizotrons. Agron. J. 76:1015-1017.

Van Straalen, N. M., and P. C. Rijninks. 1982. The efficiency of Tullgren apparatus with respect to interpreting seasonal changes in age structure of soil arthropod populations. Pedobiol. 24:197-209.

Wallwork, J. A. 1970. _Ecology of soil animals_. London: McGraw-Hill.

Warnock, A. J., A. H. Fitter, and M. B. Usher. 1982. The influence of a springtail _F. candida_ (Insecta, Collembola) on the mycorrhizal associations of leek _Allium porrum_ and the vascular arbuscular mycorrhizal endophyte _Glomus fasciculatus_. _New Phytol_. 90:285-292.

E. J. Kladivko
H. J. Timmenga

8 Earthworms and Agricultural Management

Earthworms can have a major impact on the rhizosphere by their feeding, casting, and burrowing activities. Although earthworms are relatively large compared with the average diameter of roots of many agricultural crops, they can significantly alter the root rhizosphere as well as bulk soil properties. These impacts have often been overlooked in modern rowcrop agriculture in the United States, but greater interest in reduced tillage systems and concern about agricultural sustainability have prompted researchers and farmers to again consider the potential role of the earthworm. The excellent books by Lee (1985) and Edwards and Lofty (1977) discuss earthworm biology and ecology in great detail, and readers should consult them for more information.

The objectives of this chapter are to summarize the major effects of earthworms on soil properties and plant growth and the effects of agricultural management practices on earthworm populations. We conclude by discussing the potential importance of earthworms in conservation tillage systems, management strategies that may enhance earthworm activity, and areas that need additional applied research.

BASIC ECOLOGY OF EARTHWORMS

Most references to earthworms in the scientific literature deal with the family Lumbricidae, but the ecological

This chapter was originally printed as Journal Paper no. 11,663 of the Purdue University Agricultural Experiment Station, West Lafayette, Indiana.

strategies discussed here apply to other families as well.
Although Lumbricidae may have been native to North America
as well as Europe, they probably did not survive the
Quaternary glaciations. Lumbricidae were reintroduced to
much of North America by European settlers during the past
400 years or by migration from unglaciated areas in the
south (Lee 1985). This suggests that there may be areas in
the glaciated portions of North America that lack
particular species of earthworms, because they have not
been introduced. Some unglaciated areas of the West Coast
may support indigenous non-Lumbricid species (Spiers et al.
1986).

Different species of earthworms occupy different nich-
es and are often grouped by their feeding or burrowing habits. Bouché (1977) proposed the following classification,
based on morphological and ecological characteristics, for
the European Lumbricids: epigées are litter dwellers and
feeders, endogées live and feed in the mineral soil, and
anéciques are deep-burrowing worms that live in the mineral
soil but typically feed on dead leaves at the soil surface.
We will discuss the activities of endogées and anéciques,
because only those species will have a major impact on
soils in agricultural systems. It should be noted that
feeding and burrowing habits of a species also depend on
food availability, agricultural management, and edaphic conditions (Lee 1985). The morphological-ecological classifi-
cations are, therefore, a general guide to ecological
strategies.

While lacking permanent burrows, the endogées generally burrow through the topsoil, ingesting mineral soil and
organic materials but preferentially selecting organic materials. Different species ingest organic matter in different stages of decomposition. Lumbricus rubellus, for example, ingests relatively undecomposed organic matter, while
Allolobophora caliginosa feeds on well-decomposed material
(Piearce 1978). In contrast the anéciques such as L. terrestris (commonly called "nightcrawlers") make permanent,
nearly vertical burrows into the subsoil but feed on litter
at the surface. They often pull leaves down into their burrows for aging and later feeding. These two major group-
ings of earthworms can have quite different effects on
soils and plant growth and are affected differently by agricultural management.

Earthworms are saprovores, deriving their energy primarily from ingesting dead plant materials. They also obtain nutrition from microorganisms ingested either preferentially or along with organic and mineral matter (Satchell

1983). Although they ingest large quantities of soil and organic matter, they assimilate only a small amount of the carbon ingested, typically one to three percent (Bolton and Phillipson 1976; Uvarov 1982).

THE EFFECTS OF EARTHWORMS ON SOIL PROPERTIES AND PROCESSES

Channels

Earthworms create burrows typically ranging in diameter from 1 to 10 mm. These burrows can conduct water and air rapidly into the soil and provide low-resistance pathways for root growth. The influence of earthworm burrows on water flow is greater on soils that have a low soil matrix permeability. Field and greenhouse experiments have shown large increases in water infiltration rates due to the presence of earthworm burrows (Hopp and Slater 1948; Ehlers 1975; Edwards et al. 1979; Bouma et al. 1982; Kladivko et al. 1986; Zachmann et al. 1987; Kemper et al. 1987). Recent evidence suggests that water will infiltrate in some earthworm burrows even when rainfall amounts are relatively low, especially if rainfall is intense in the initial part of the storm and the soil surface is initially dry (Edwards et al. 1989). Infiltration in macropores produces a different soil water content distribution with depth than in soils without burrows. The role of earthworm burrows and other macropores in movement of water and chemicals through the soil is the subject of much current research.

The rate and location of burrow formation is affected by availability and location of the food supply as well as by soil water content, temperature, bulk density, and penetration resistance (Lavelle 1975; Bolton and Phillipson 1976; Timmenga 1987). Zachmann, Linden, and Clapp (1987) added A. tuberculata or L. rubellus to no-till plots with surface residues and to tilled plots with incorporated residues in Minnesota. Over a six-week period in early summer, both species formed twice as many burrows open to the soil surface on no-till as on tilled plots (Table 8.1). Because the major food supply was at the soil surface in no-till but was buried in the tilled plots, feeding and burrowing activity near the surface was greater with no-till than with tillage. The surface residue also provides a suitable habitat for the earthworms by insulating the soil from

TABLE 8.1. The number of burrows open to the soil surface after six weeks of earthworm activity in Minnesota, as affected by tillage treatment and species.

Tillage treatment	Species	Number of burrows (SE)[a]
Tilled[b]/ incorporated residue	A. tuberculata	14.5 (0.6)
	L. rubellus	15.0 (0.8)
No-till/ surface residue	A. tuberculata	36.5 (3.7)
	L. rubellus	47.5 (4.4)

[a]Measured in 0.06 m^2 subplots inoculated with 13 earthworms each.
[b]Tilled with rototiller to 10-cm depth.
Source: From Zachmann, Linden, and Clapp (1987).

rapid drying or heating in early summer.

Several researchers have counted earthworm burrows or other biopores under different agricultural management systems (Ehlers 1975; Boone et al. 1976; Gantzer and Blake 1978; Barnes and Ellis 1979; Edwards and Norton 1986) and have assessed their continuity from the soil surface to the subsoil with the use of dyes or tracers (Ehlers 1975; Douglas et al. 1980; Germann et al. 1984; Zachmann et al. 1987; Shipitalo and Edwards 1987; Heard et al. 1988). As many as 1,700 biopores (earthworm and old root channels) greater than 1 mm in diameter have been found in one square meter of soil area at a depth of 30 to 38 cm (Gantzer and Blake 1978). Generally, channels are more numerous and continuous in no-till than in tilled fields because the burrow openings in the topsoil are not destroyed each year by tillage. Burrows in the subsoil may persist for many years after their inhabitants have left, even if the burrows are destroyed in the topsoil. Shipitalo and Edwards (1987) found large L. terrestris burrows in the subsoil that had been abandoned for at least twenty-four years. Regardless of whether they extend to the surface, burrows can aid root development (Wang et al. 1986) and will be discussed in more detail in a later section.

Casts

Organic materials and mineral soil are mixed intimately during passage through the earthworm gut and are excreted as casts. Casts are deposited at the soil surface, in subsurface cavities, or along burrow walls, depending on the species and soil conditions. The amount of casts produced depends on food source and environmental conditions in the soil. Annual surface cast production in agricultural systems ranges from 4 to 90 Mg ha^{-1} in temperate regions (Edwards and Lofty 1977; Lee 1985).

Earthworm casts influence many physical, chemical, and biological conditions of the soil. Casts generally contain more organic matter and available nutrients than the surrounding soil because of the earthworm's selective ingestion of organic materials and the mineralization of those materials during passage through the gut (Lee 1985; Timmenga 1987). Casts often contain higher concentrations of exchangeable calcium, magnesium, and potassium (Lunt and Jacobson 1944) and extractable phosphorus (Sharpley and Syers 1976, 1977) than the surrounding soil. Higher phosphatase (Satchell and Martin 1984) and urease (Syers et al. 1979) activity has also been found in casts. The pH of casts is generally closer to neutral than that of the parent soil (Edwards and Lofty 1977). Microbial populations are larger in casts than in the surrounding soil (Parle 1963) due to proliferation of microbes during passage through the gut and in the cast after deposition. Although total microbial populations are greater, there is evidence that cells of some microorganisms are destroyed during their passage through the earthworm gut (Edwards and Lofty 1977; Rouelle 1983). Earthworms may act as vectors for bacteria and fungi, including plant pathogens, by ingesting cells or spores in one place and depositing them in casts elsewhere (Rouelle 1983; Shaw and Pawluk 1986).

Casts have generally been found to increase the water-stable aggregation of soils (Hopp and Hopkins 1946a; Lal and Akinremi 1983; Kladivko et al. 1986). Under some pasture conditions, however, surface casting by earthworms caused an increase in sediment losses by erosion (Darwin 1881; Sharpley et al. 1979). These greater sediment losses may be due in part to exposing soil material (casts) in an area otherwise covered with growing plants and protected

from erosion. Shipitalo and Protz (1988) found that freshly deposited moist casts are not very stable, but aging and
drying processes increase their water stability.   Cast
stability, as measured by clay dispersibility in the casts,
also varied with the earthworm species and the food source.
    Earthworm burrows, especially the "permanent" burrows
of anéciques, are stabilized by castings deposited along
burrow walls and by mucus excreted by earthworms (Piearce
1981).   Burrow walls can thus be viewed as stable, well-
aerated soil zones containing higher concentrations of organic matter, available nutrients, and microorganisms than
the surrounding soil. Bouché (1975) discussed the "drilosphere," a zone 2 mm thick around burrow walls, as a zone
of increased microbial activity due to these conditions.
The importance of this zone for increased root growth and
nutrient uptake, especially in less fertile soils, needs
further investigation.

Mixing

    Earthworms mix organic debris and applied inorganic
materials (fertilizer phosphorus, lime, pesticides) into
the soil and may be very important in cycling organic matter and nutrients in forest, pasture, and no-till ecosystems.   Much of the research on soil mixing by earthworms
has been conducted in pastures, forests, orchards, or drastically disturbed lands under reclamation.   The importance
of soil mixing by earthworms in nontilled rowcrop agriculture needs further investigation.
    Stockdill (1982) has done much work on the effect of
earthworms on pasture productivity in New Zealand. Earthworms incorporate thatch materials as well as applied lime,
nutrients, and insecticides. Some species, such as A. caliginosa and L. rubellus, tend to mix the lime laterally,
while others such as A. longa mix it vertically (Springett
1983).   The incorporation and intimate mixing of phosphate
rock pellets with soil by earthworms indirectly increases
the availability of phosphorus from the rock (Mackay et al.
1982).
    Dietz and Bottner (1981) placed litter labelled with
^{14}C on the surface of a grass land soil and used autoradiography to follow the movement of the ^{14}C into the profile.
Most of the decomposition products were transported through
the profile by water, while a small portion was mixed into
the soil by earthworms. The introduction of L. terrestris
to coal spoil banks in Ohio (Vimmerstedt and Finney 1973)

and reclaimed peat in Ireland (Curry and Bolger 1984)
resulted in greater litter incorporation than in similar
disturbed lands without the earthworms.

Earthworms--<u>Allolobophora tuberculata</u>, <u>Dendrobaena</u>
<u>octaedra</u>, <u>L. festivus</u>, and <u>L. terrestris</u>--invaded a New
Brunswick mixed forest and dramatically changed the soil
profile. The transformation from a typical podzol to a
profile with an apparent Ah horizon required only four
years (Langmaid 1964). Introductions of earthworms <u>A. ca-</u>
<u>liginosa</u> and <u>L. terrestris</u> into newly reclaimed polders in
the Netherlands improved soil structure and decreased
thatch accumulations (Hoogerkamp et al. 1983). When overuse of copper-containing pesticides killed the earthworms
in an orchard soil in the Netherlands, soil structure rapidly deteriorated and a thatch layer developed on top of the
mineral soil (van Rhee 1963).

Timmenga (1987) measured residue mixing and cast distribution in a soil column experiment. Silty clay loam
soil was packed into columns (10-cm diameter, 35-cm depth)
to a bulk density of 1.0 g cm^{-3}. A water table was maintained 36 cm from the surface. Plant parts from red clover
(<u>Trifolium pratense</u>) were added as a food source. The root
material was placed at the 5-cm depth, while the shoot material (mostly leaves) was placed on the soil surface. Five
earthworms (<u>L. rubellus</u>) were added to each soil column except the controls. Columns were capped, and evolved CO_2
was trapped to monitor decomposition. After thirty days,
columns were frozen and sectioned. Cast material and plant
remains were collected, and observations of burrows were
made at depth increments throughout the column.

The earthworms did not have a noticeable effect on the
rate of CO_2 evolution, probably because they do not assimilate organic matter in large quantities. In columns without earthworms, losses in dry weight were 56 and 6.9 percent for shoot and root materials, respectively. In the
presence of earthworms, shoot material was almost completely removed from the surface of the soil, while most of the
root material was not moved. Horizontal burrows were found
near where the roots were buried, indicating that earthworms were active at that depth, but only slight amounts of
the root material itself were consumed by the worms. This
suggests that the root material lacked palatability or perhaps a microbial population. In the 0- to 5-cm layer, the
burrow walls (all cast material) were thick and easy to
recognize and recover because of their dark olive green
color, rounded shape, and high fiber content. Some of the
burrows were blocked with casts. Along the sides of the

columns, the worms had created reinforced burrows made of large amounts of casts. In the 5- to 10-cm layer, the burrow walls were not very thick but could easily be removed by peeling the burrow wall material from the surrounding soil. Most burrows below 10 cm occurred along the sides of the column, where the soil was probably looser, better aerated, and more penetrable than soil inside the column. These burrows along the sides of the column had very thin walls with virtually no recoverable casts, and some burrows showed only slightly smeared tracks. Burrows were found in the soil close to the water table, suggesting that a high soil water potential per se does not restrict the burrowing depth of L. rubellus.

Few casts were found in the 20+ cm layer, and cast weight was negatively correlated with depth in the column. The bulk of the cast material was found on the surface (16%), and in the 0-5 (46%), the 5-10 (22%), and the 10-15 (12%) cm layers. The average dry weight of casts recovered from each column was 34.28 g (SE 0.91). The average dry weight of the earthworms, measured in a subsample of the population before the experiment, was 0.094 g worm^{-1} (SE 0.0066). The egestion rate of L. rubellus in the experiment was estimated at 2.43 g dry cast weight g^{-1} dry worm weight day^{-1}, calculated for columns from which five worms were recovered.

THE EFFECTS OF EARTHWORMS ON PLANT GROWTH AND YIELD

Root Growth

Root growth can be increased, especially in subsoils or moderately dense soils, by the presence of earthworms or their burrows (Figure 8.1). Edwards and Lofty (1978, 1980) added earthworms to small field plots and to intact soil columns taken from direct-drilled (no-till) plots and found increased root growth compared to the noninoculated treatments (Table 8.2). The root distributions were closely correlated with the zones of activity of the particular earthworms added. Deep-burrowing (anécique) species (L. terrestris, A. longa) promoted deeper root growth, and the more shallow-dwelling (endogeic) species (A. caliginosa, A. chlorotica) promoted shallow root growth. Tests with artificially formed burrows compared with earthworm burrows suggested that the improved root growth resulted from both

Figure 8.1. Corn roots growing in earthworm burrows at a depth of 60 cm.

the physical ease of root growth in preformed channels and the greater availability of nutrients in the burrow walls. Banded applications of nitrogen and phosphorus to soils usually result in root proliferation in those bands (Duncan and Ohlrogge 1958). Roots would tend to proliferate in earthworm burrows due to available phosphorus from cast material and available nitrogen from urine or mucus excretions.

Root proliferation in earthworm burrows in the subsoil has been observed by several researchers. Ehlers et al. (1983) found that oat roots were detectable within biopores below a 10-cm depth in untilled loess soil and below a 15-cm depth in tilled soil. The percentage of roots found in

TABLE 8.2. Root growth of barley grown in undisturbed profiles of silty clay loam soil, as affected by earthworm additions and tillage treatment.

Tillage treatment	Earthworm species	Root weight[a] (g)
Direct drill	None	0.39
Direct drill	L. terrestris[b]	0.79
Direct drill	A. long[ab]	0.78
Direct drill	A. caliginosa and A. chlorotica[c]	0.62
Plow	None	1.04

[a]Standard error = 0.13g.
[b]Five mature earthworms added per box.
[c]A total of ten mature earthworms added per box.

Source: From Edwards and Lofty (1978).

biopores increased with soil depth, and below 50- cm nearly all the roots were in biopores. On a silt loam soil in Illinois, Wang, Hesketh, and Wooley (1986) found soybean roots entering burrows at depths of 30 to 45 cm. The roots then grew within the burrows down to the point where the burrows ended, which was as deep as 130 cm. If a taproot did not encounter a burrow by the time it reached the 45-cm depth, the root tip died.

In polder soils of the Netherlands, large-scale inoculations of earthworms were made to study the effects of earthworm activity on soil properties and plant growth. Van Rhee (1977) found greater total root growth and a higher ratio of thin to thick roots (less than 0.5 mm compared with 0.5-5 mm) of young apple trees in the earthworm plots, probably due to increased soil aggregation. Only very small differences in fruit yields were observed, however. A deep-burrowing species of earthworm (A. longa) was introduced to some New Zealand pasture soils that were already inhabited by shallow-dwelling earthworms. The inoculated plots had increased soil porosity at the 10- to 20-cm depth and greater root development in the 15- to 20-cm depth than the control plots (Springett 1985). Pasture growth was increased during some seasons.

Shoot Growth and Yield

Although root growth increases under some conditions as a result of earthworm activity, this may or may not result in increased shoot growth or crop yield in any given year. Extensive research on earthworms in pasture soils has consistently shown increased pasture production in the presence of earthworms. Lee (1985) discusses the history of pasture improvement by earthworm introductions in New Zealand, and an example of Stockdill's work is presented in Table 8.3. Earthworms were introduced in 1949 at various points in an established pasture. After four years, the earthworms were well established at the points of introduction and were improving early spring growth of the pasture. Once the earthworms were established, they advanced about 10 m per year across the field. Measurements of pasture production were made in the 1965-66 growing season in three different zones of the field: zone 1, where earthworms had been established for ten years; zone 2, where earthworms were just entering during 1965-66; and zone 3, where there were still no worms. The 30 percent higher yield in zone 1 over zone 3 was attributable to incorporation of thatch material, improvement in porosity and infiltration, and mixing

TABLE 8.3 Pasture yields in New Zealand, as affected by earthworm (A. caliginosa) introduction.

	Dry matter yield		
	Zone 3 (no worms)	Zone 2 (worms entering)	Zone 1 (worms established)
	----------Mg ha^{-1}----------		
Spring 1965	3.26	3.61	4.10
Summer 65/66	3.76	4.12	4.78
Autumn 1966	2.38	3.45	3.22
Total 65/66	9.40	11.18	12.10

Source: From Stockdill (1982).

of lime and nutrients into the soil (Stockdill 1982). Earthworms have also increased pasture production in polder soils in the Netherlands (Hoogerkamp et al. 1983).

The growth of various crops as affected by earthworms has been measured in many pot experiments. Hopp and Slater (1949) conducted a series of experiments with soybeans, lima beans, millet, wheat, and hay grown in greenhouse pots, soil frames, or large barrels with either no earthworms, dead earthworms, or live earthworms added. All crops had greater yields with dead earthworms compared with no earthworms, due to chemicals released from the decaying earthworms. Additional yield increases occurred in the presence of live earthworms, on those soils that were initially poorly structured or artificially puddled, presumably due to improved soil structure caused by earthworm activity. With large populations of earthworms, van Rhee (1965) found increases in shoot dry matter production of wheat, grass, and clover, but decreases in pea yields. Atlavinyte and Zimkuviene (1985) planted barley in soil pots at three different bulk densities, with or without earthworms and N-P-K fertilizer. Earthworms accelerated germination and heading and increased grain yield in the soils at higher bulk densities (1.3 and 1.5 g cm^{-3}). During a twelve-month study, A. caliginosa enhanced early growth and increased shoot dry matter production of perennial ryegrass on a silt loam soil from which the topsoil (0 to 15 cm) had been stripped (McColl, Hart, and Cook 1982).

Several researchers have suggested that earthworms stimulate plant growth by the production of metabolites similar to plant growth hormones or other biologically active substances. Graff and Makeschin (1980) found greater shoot dry matter production of ryegrass, both in pots in which earthworms were removed after eleven days of activity and in pots watered with eluates from the earthworm pots. They suggested that earthworms released some yield-influencing subtances into the pots. Edwards et al. (1985) reported that earthworm casts enhanced plant growth and germination above that due to the enriched nutrient content alone. Springett and Syers (1979) observed rye-grass roots growing into surface casts of L. rubellus but not A. caliginosa. They suggested that the negatively geotropic growth may be due to some auxin-like substance in the casts and not solely to gradients of available phosphorus. Increased concentrations of vitamin B12 in the soil have been attributed to earthworm activity (Atlavinyte and Daciulyte 1969) or to the stimulation of microbial activity associated with earthworms.

EFFECT OF AGRICULTURAL MANAGEMENT ON EARTHWORMS

Tillage and Cropping

Agricultural management systems affect earthworm populations and activity by affecting the amount, quality, and location of food for the earthworms and by changing the physical and chemical properties of the soil. Earthworm populations are larger under permanent pasture than under continuously cropped land (Evans and Guild 1948; Hopp and Hopkins 1946b; Barley 1959) primarily because of the much greater supply of organic materials under pasture. The loss of a surface insulation layer may also contribute to the decline in earthworm populations with cultivation, especially in harsh climates (very cold or very dry). The tillage operation itself will kill some earthworms, but this generally represents only a minor contribution to the population declines on tilled land.

After a pasture is plowed and cultivated, earthworm populations remain stable for some time because a large food supply is still available. With continued cultivation, the populations decline (Evans and Guild 1948) and the species distribution changes. Farmers are interested in knowing how quickly earthworm populations will change in response to changes in management practices. Because populations will change as land goes into government "set-aside" programs or into crop rotations with hay crops, for example, some practical management options for increasing populations may be available to farmers. It appears hat several years of pasture is sufficient to rebuild large earthworm populations, but that plowing and cultivation will reduce those populations again within several years. Barley (1959) compared earthworm populations in a two-year rotation (wheat-fallow), a four-year rotation (pasture-pasture-fallow-wheat), and permanent pasture in Australia. The two-year rotation had very few earthworms at any sampling time. In the four-year rotation, earthworm biomass was nearly as great after the two years of pasture as in permanent pasture, but dropped to about one-fourth the biomass after the fallow and wheat years. Possibly the use of no-till rather than plowing after the pasture years would slow the rate of decline of earthworm populations.

The influence of different tillage practices on earthworm populations and activity has been studied primarily in Europe, where climatic conditions are very different than in many parts of the continental United States. Direct drilling of wheat or barley crops generally resulted in greater earthworm populations than did plowing (Barnes and Ellis 1979; Gerard and Hay 1979; Edwards and Lofty 1982; Clutterbuck and Hodgson 1984). Deep-burrowing (anécique) species were affected more by tillage than the shallow dwellers. The plow destroys the natural feeding habitat of the anécique by burying the residues below the soil surface. Tillage destroys the upper portion of the permanent burrows of anéciques, which must then be reconstructed. Replacement of individuals killed during the tillage operation itself is slower for anéciques than for shallow dwellers because of lower reproductive rates. But even the shallow dwelling earthworms had larger populations with direct drilling than with plowing, probably because of less destruction by cultivation operations, organic materials being available over a longer period of time, and insulation of the soil surface. A surface mulch or protective layer was important in maintaining earthworm populations through the winter in north-central and mid-Atlantic regions of the United States where soils freeze (Hopp 1947; Slater and Hopp 1947).

In Indiana, Mackay and Kladivko (1985) found twice as many earthworms under no-till as under moldboard plowed plots growing continuous soybeans, but found no differences between the tillage systems in continuous corn (Table 8.4). Populations were very low in the continuous corn plots sampled. Mackay and Kladivko suggested that the wider C:N ratio of corn residue compared with soybean residue and the application of anhydrous NH3 and the soil insecticide terbufos in the corn plots may have limited the populations. De St. Remy and Daynard (1982) also found no consistent increases in the numbers or biomass of earthworms under no-till corn in Ontario, Canada. However, Lal (1976) found an average five-fold increase in the numbers of earthworm casts under no-till compared with disc plowing under both continuous soybeans and corn in Nigeria. In Georgia, House and Parmelee (1985) found higher earthworm populations with no-till than with moldboard plowing in a sorghum-soybean rotation that included winter cover crops.

Organic matter additions, in the form of manures or sewage sludges, greatly increase earthworm populations (Edwards and Lofty 1982; Standen 1984; Mackay and Kladivko 1985). Some sludges may have a brief initial detrimental effect (Edwards and Lofty 1982), presumably because of high ammonium concentration. In general, the greater the amount of organic matter added to soils, the higher the earthworm populations will be. Nitrogen content and other quality

TABLE 8.4. Earthworm populations as affected by crop and tillage practice, autumn 1983, in Indiana.

Crop	Tillage	Earthworms no. m^{-2} (SD)	
Corn	Plow	8	(8)
Corn	No-till	16	(8)
Soybeans	Plow	62	(23)
Soybeans	No-till	141	(9)
Clover/ryegrass	Pasture	470	(5)
Clover/ryegrass	Pasture[a]	1,298	(75)

[a]This site also received large applications (more than 10 Mg ha^{-1} y^{-1}) of dairy cow manure from the barnyard.

Source: From Mackay and Kladivko (1985).

components of the organic material also affect earthworm growth. For example, Shipitalo, Protz, and Tomlin (1988) offered excess food to L. terrestris and L. rubellus in soil pots in the lab. They estimated food ingestion as the rate of disappearance of food from the soil surface. They found that both species ingested the most food and gained the most weight when offered alfalfa or red clover leaves and had intermediate ingestion and weight gain with corn leaves, but rejected bromegrass leaves as food and therefore lost weight.

Fertilizers and Pesticides

There are conflicting reports on the effects of inorganic fertilizers on earthworm populations in the field. Fertilizer additions increase plant growth and therefore increase the rate of food supply--dead roots and residues-- available to earthworms. The acidifying effect of some inorganic fertilizers may account for population reductions that have been observed (Lee 1985). Earthworm species differ in their optimal pH range (Edwards and Lofty 1977), but many species tolerate a soil pH between 5 and 7.

Earthworm populations have been shown to increase with increasing nitrogen rates up to 192 kg N ha^{-1} applied as $Ca(NO_3)_2$ (Edwards and Lofty 1982). When farmyard manure plus $(NH_4)_2SO_4$ or $NaNO_3$ was added to cropland, Edwards and Lofty (1982) found larger earthworm populations with $(NH_4)_2SO_4$, which conflicted with earlier research. Potter, Bridges, and Gordon (1985) found earthworm populations decreased with increased NH_4NO_3 rates and were correlated with decreased soil and thatch pH over the 6.0-4.5 pH range on turfgrass. Gerard and Hay (1979) found earthworm populations increased with nitrogen additions up to 50 or 100 kg N ha^{-1} as $Ca(NO_3)_2$, and stayed the same or decreased slightly at 150 kg N ha^{-1}.

There are no published reports of field experiments to assess earthworm response to the more common nitrogenous fertilizer forms used for corn production in the United States--anhydrous ammonia, urea, and urea-ammonium nitrate (UAN) solution. Fertilizer placement--whether surface broadcast or banded or injected--and timing may also affect earthworm response. Ammonia itself is toxic to earthworms, but the overall effect on field populations of injecting anhydrous ammonia is not known. Because the location of earthworms in the profile and their age distribution change throughout the season, mortality might be lessened by prudent timing of chemical application (preplant vs. side-dressed ammonia in corn, for example).

Data on the effect of different pesticides on earthworm populations or activity were compiled by Edwards (1980) and by Lee (1985). Many insecticides and fungicides are toxic to earthworms, but many herbicides are relatively harmless. Again, the placement of the pesticide (broadcast vs. banded over the seed row) may affect earthworm mortality by affecting the proportion of the population exposed to the chemical.

SUMMARY AND IMPLICATIONS

Earthworms are an important component in the soil ecological system. Earthworms form burrows that can improve water infiltration, drainage, and aeration of the soil profile, as well as increasing the extent and intensity of rooting. Following aging or drying, cast material is very stable and generally improves soil tilth, although fresh, moist casts are usually very unstable. Cast material and mucus secretions tend to stabilize earthworm burrows while providing a nutrient-rich lining of the burrow that may further increase root proliferation. Earthworms mix organic materials and surface-applied fertilizers and lime into the soil, both actively by burrowing and feeding and passively as a result of water transport of materials into their burrows. Thatch and root mat layers have been incorporated into the soil, enhancing root development and water infiltration. Incorporation of surface residues by earthworms also speeds their subsequent degradation by other soil organisms.

The improvements in soil properties brought about by earthworm activity would be expected to increase root growth or plant yield primarily under some type of "stress" condition. Physical stresses--compaction, dense subsoils, crusting, low water infiltration or water availability, poor aeration--and chemical stresses such as low fertility are quite common and offer opportunities to explore the potential importance of earthworms in modern rowcrop agriculture. The use of earthworms to help reclaim coal spoils (Vimmerstedt and Finney 1973) and peat bogs (Curry and Bolger 1984) are examples of the importance of earthworm activity in amelioration of stressful plant habitats.

Earthworms may be more important in no-till and other reduced tillage systems than in conventional moldboard plowed systems, not only because populations are higher under no-till but also because mechanical mixing and loosening by tillage is diminished. When residues are not incorporated by a plow, they can be slowly incorporated by earthworms and other organisms. In the absence of mechanical loosening of the soil, earthworm burrows and casts may become more important for good root development. Increased earthworm activity may partially substitute for the beneficial effects of the moldboard plow. The topsoil will not be as "well mixed" by earthworms as by a plow, and different management strategies may be needed to fully utilize the potential of the biological populations.

Several agricultural management strategies can be used to increase earthworm populations and activity. The particular strategy employed will depend somewhat on whether anécique or endogeic earthworms are of greater interest in that system. Increasing the amount and quality of the food source available, including plant residues and manures, is probably one of the most important ways to increase populations in agricultural fields. Leaving the soil surface covered with organic materials, both as a feeding habitat for deep-burrowing species and as a protective mulch against rapid freezing or drying, will also promote earthworm populations and may prolong their active period in the spring and autumn. Growing winter cover crops on either tilled or nontilled fields may increase populations by adding more food and protecting the surface. Elimination of tillage will be especially beneficial to deep-burrowing species because their permanent burrows will not continually be destroyed. Modifications in the use of pesticides and inorganic nitrogenous fertilizers may reduce earthworm mortality, and drainage of wet soils may promote larger earthworm populations.

Additional research is needed in many areas of earthworm agroecology. Some of the questions that are important include: How significant is the "drilosphere" for root proliferation and plant nutrient uptake under different soil conditions? Can the excretion of organic metabolites or growth hormones by earthworms be important in plant growth? How much can the use of winter cover crops increase earthworm populations within a typical crop rotation? How significant would mixing activities be in no-till or other reduced tillage systems? Under what soil conditions would improvements in soil physical properties by earthworms be both large and important? Can changes in form, timing, or placement of fertilizers and pesticides reduce any adverse effects on earthworms while still fulfilling their function? Answers to these questions will help improve agricultural management systems.

REFERENCES

Atlavinyte, O., and J. Daciulyte. 1969. The effect of earthworms on the accumulation of vitamin B12 in soil. Pedobiology 9:165-170.

Atlavinyte, O., and A. Zimkuviene. 1985. The effect of earthworms on barley crop in the soil of various density. Pedobiology 28:305-310.

Barley, K. P. 1959. The influence of earthworms on soil fertility. I. Earthworm populations found in agricultural land near Adelaide. Australian Journal of Agricultural Research 10:171-178.

Barnes, B. T., and F. B. Ellis. 1979. Effects of different methods of cultivation and direct drilling and disposal of straw residues on populations of earthworms. Journal of Soil Science 30:669-679.

Bolton, P. J., and J. Phillipson. 1976. Burrowing, feeding, egestion and energy budgets of Allolobophora rosea (Savigny) (Lumbricidae). Oecologia 23:225-245.

Boone, F. R., S. Slager, R. Miedema, and R. Eleveld. 1976. Some influences of zero-tillage on the structure and stability of a fine-textured river levee soil. Netherlands Journal of Agricultural Science 24:105-119.

Bouché, M. B. 1975. Action de la faune sur états de la matiere organique dans les ecosystemes. In Biodegradationet humification, ed. K. Gilbertus, O. Reisinger, A. Mourey, and J. A. Cancela da Fonseca, 57-168. Sarruguemines, France: Pierron.

Bouché, M. B. 1977. Strategies lombriciennes. In Soil organisms as components of ecosystems, ed. U. Lohm and T. Persson. Biology Bulletin (Stockholm) 25:122-132.

Bouma, J., C. F. M. Belmans, and L. W. Dekker. 1982. Water infiltration and redistribution in a silt loam subsoil with vertical worm channels. Soil Science Society of America Journal 46:917-921.

Carter, A., J. Heinonen, and J. De Vries. 1982. Earthworms and water movement. Pedobiology 23:395-397.

Clutterbuck, B. J., and D. R. Hodgson. 1984. Direct drilling and shallow cultivation compared with ploughing for spring barley on a clay loam in northern England. Journal of Agricultural Science (Cambridge) 102:127-134.

Curry, J. P., and T. Bolger. 1984. Growth, reproduction, and litter and soil consumption by Lumbricus terrestris L. in reclaimed peat. Soil Biology and Biochemistry 16:253-257.

Darwin, C. R. 1881. The formation of vegetable mould through the action of worms, with observations on their habits. London: Murray.

De St. Remy, E. A., and T. B. Daynard. 1982. Effects of tillage methods on earthworm populations in monoculture corn. Canadian Journal of Soil Science 62:699-703.

Dietz, S., and P. Bottner. 1981. Etude par autoradiographie de l'enfouissement d'une litière marquée au 14C en milieu herbacé. Colloques internationaux du Centre National des Recherches Scientifiques 303:125-132.

Douglas, J. T., M. J. Goss, and D. Hill. 1980. Measurements of pore characteristics in a clay soil under ploughing and direct drilling, including use of a radioactive tracer (144Ce) technique. Soil Tillage Research 1:11-18.

Duncan, W. G., and A. J. Ohlrogge. 1958. Principles of nutrient uptake from fertilizer bands. II. Root development in the band. Agronomy Journal 50:605-608.

Edwards, C. A. 1980. Interactions between agricultural practice and earthworms. In Soil biology as related to land use practice, ed. D. Dindal. Proceedings of the seventh international colloquium of soil zoology. EPA-560/13-80-038.

Edwards, C. A., I. Burrows, K. E. Fletcher, and B. A. Jones. 1985. The use of earthworms for composting farm wastes. In Composting of agricultural and other wastes, ed. J. K. R. Gasser, 229-241. Proc. CEC seminar, Oxford, March 19-20, 1984. London: Elsevier.

Edwards, C. A., and J. R. Lofty. 1977. Biology of earthworms. London: Chapman and Hall.

Edwards, C. A., and J. R. Lofty. 1978. The influence of arthropods and earthworms upon root growth of direct drilled cereals. Journal of Applied Ecology 15:789-795.

Edwards, C. A., and J. R. Lofty. 1980. Effects of earthworm inoculation upon the root growth of direct drilled cereals. Journal of Applied Ecology 17:533-543.

Edwards, C. A., and J. R. Lofty. 1982. Nitrogenous fertilizers and earthworm populations in agricultural soils. Soil Biology and Biochemistry 14:515-521.

Edwards, W. M., R. R. van der Ploeg, and W. Ehlers. 1979. A numerical study of the effects of noncapillary-sized pores upon infiltration. Soil Science Society of America Journal 43:851-856.

Edwards, W. M., and L. D. Norton. 1986. Effect of macropores on infiltration into non-tilled soil. _Transactions of the Thirteenth Congress of the International Soil Science Society_ 5:47-48.

Edwards, W. M., M. J. Shipitalo, L. B. Owens, and L. D. Norton. 1989. Water and nitrate movement in earthworm burrows within long-term no-till cornfields. _Journal of Soil and Water Conservation_ 44:240-243.

Ehlers, W. 1975. Observations on earthworm channels and infiltration on tilled and untilled loess soil. _Soil Science_ 119:242-249.

Ehlers, W., U. Kopke, F. Hesse, and W. Bohm. 1983. Penetration resistance and root growth of oats in tilled and untilled loess soil. _Soil and Tillage Research_ 3:261-275.

Evans, A. C., and W. J. McL. Guild. 1948. Studies on the relationships between earthworms and soil fertility. V. Field populations. _Annals of Applied Biology_ 35:485-493.

Gantzer, C. J., and G. R. Blake. 1978. Physical characteristics of Le Sueur clay loam soil following no-till and conventional tillage. _Agronomy Journal_ 70:853-857.

Gerard, B. M., and R. K. M. Hay. 1979. The effect on earthworms of ploughing, tined cultivation, direct drilling and nitrogen in a barley monoculture system. _Journal of Agricultural Science_ (Cambridge) 93:147-155.

Germann, P. F., W. M. Edwards, and L. B. Owens. 1984. Profiles of bromide and increased soil moisture after infiltration into soils with macropores. _Soil Science Society of America Journal_ 48:237-244.

Graff, O., and Makeschin, F. 1980. Beeinflussung des Ertrags von Weidelgras (_Lolium multiflorum_) durch Ausscheidungen von Regenwurmen dreier verschiedener Arten. _Pedobiology_ 20:176-180.

Heard, J. R., E. J. Kladivko, and J. V. Mannering. 1988. Soil macroporosity, hydraulic conductivity and air permeability of silty soils under long-term conservation tillage in Indiana. _Soil and Tillage Research_ 11:1-18.

Hoogerkamp, M., H. Rogaar, and H. J. P. Eijsackers. 1983. Effect of earthworms on grassland on recently reclaimed polder soils in the Netherlands. In _Earthworm ecology_, ed. J. E. Satchell, 85-106. London: Chapman and Hall.

Hopp, H. 1947. The ecology of earthworms in cropland. _Soil Science Society of America Proceedings_ 12:503-507.

Hopp, H., and H. T. Hopkins. 1946a. Earthworms as a factor in the formation of water-stable aggregates. _Journal of Soil and Water Conservation_ 1:11-13.

Hopp, H., and H. T. Hopkins. 1946b. The effect of cropping systems on the winter population of earthworms. _Journal of Soil and Water Conservation_ 1:85-88.

Hopp, H., and C. S. Slater. 1948. Influence of earthworms on soil productivity. _Soil Science_ 66:421-428.

Hopp, H., and C. S. Slater. 1949. The effect of earthworms on the productivity of agricultural soil. _Journal of Agricultural Research_ 78:325-339.

House, G. J., and R. W. Parmelee. 1985. Comparison of soil arthropods and earthworms from conventional and no-tillage agroecosystems. _Soil and Tillage Research_ 5:351-360.

Kemper, W. D., T. J. Trout, A. Segeren, and M. Bullock. 1987. Worms and water. _Journal of Soil and Water Conservation_ 42:401-404.

Kladivko, E. J., A. D. Mackay, and J. M. Bradford. 1986. Earthworms as a factor in the reduction of soil crusting. _Soil Science Society of America Journal_ 50:191-196.

Lal, R. 1976. No-tillage effects on soil properties under different crops in western Nigeria. _Soil Science Society of America Journal_ 40:762-768.

Lal, R., and O. O. Akinremi. 1983. Physical properties of earthworm casts and surface soil as influenced by management. _Soil Science_ 135:114-122.

Langmaid, K. K. 1964. Some effects of earthworm invasion in virgin podzols. _Canadian Journal of Soil Science_ 44:34-37.

Lavelle, P. 1975. Consommation annuelle de terre par une population naturelle de vers de terre (_Millsonia anomala_ Omodeo, Acanthodrilidae-Oligochètes) dans la savane de Lamto (Cote d'Ivore). _Revue d' Ecologie et de Biologie du Sol_ 12:11-24.

Lee, K. E. 1985. _Earthworms: Their ecology and relationships with soils and land use._ Sydney, Australia: CSIRO.

Lunt, H. A., and Jacobson, G. M. 1944. The chemical composition of earthworm casts. _Soil Science_ 58:367-376.

McColl, H. P., P. B. S. Hart, and F. J. Cook. 1982. Influence of earthworms on some chemical and physical properties, and the growth of ryegrass on a soil after topsoil stripping--a pot experiment. _New Zealand Journal of Agricultural Research_ 25:229-237.

Mackay, A. D., and E. J. Kladivko. 1985. Earthworms and the rate of breakdown of soybean and maize residues in soil. Soil Biology and Biochemistry 17:851-857.

Mackay, A. D., J. K. Syers, J. A. Springett, and P. E. M. Greg. 1982. Plant availability of phosphorus in superphosphate and a phosphate rock as influenced by earthworms. Soil Biology and Biochemistry 14:281-287.

Parle, J. N. 1963. A microbiological study of earthworm casts. Journal of General Microbiology 31:13-22.

Piearce, T. G. 1978. Gut content of some lumbricid earthworms. Pedobiology 18:153-157.

Piearce, T. G. 1981. Losses of surface fluids from lumbricid earthworms. Pedobiology 21:417-426.

Potter, D. A., B. L. Bridges, and F. C. Gordon. 1985. Effect of N fertilization on earthworm and microarthropod populations in Kentucky bluegrass turf. Agronomy Journal 77:367-372.

Rouelle, J. 1983. Introduction of amoebae and Rhizobium japonicum into the gut of Eisenia foetida (Sav.) and Lumbricus terrestris (L). In Earthworm ecology, ed. J. E. Satchell, 375-382. London: Chapman and Hall.

Satchell, J. E. 1983. Earthworm microbiology. In Earthworm ecology, ed. J. E. Satchell, 351-364. London: Chapman and Hall.

Satchell, J. E., and K. Martin. 1984. Phosphatase activity in earthworm faeces. Soil Biology and Biochemistry 16:191-194.

Sharpley, A. N., and J. K. Syers. 1976. Potential role of earthworm casts for the phosphorus enrichment of runoff waters. Soil Biology and Biochemistry 8:341-346.

Sharpley, A. N., and J. K. Syers. 1977. Seasonal variation in casting activity and in the amounts and release to solution of phosphorus forms in earthworm casts. Soil Biology and Biochemistry 9:227-231.

Sharpley, A. N., J. K. Syers, and J. A. Springett. 1979. Effect of surface-casting earthworms on the transport of phosphorus and nitrogen in surface runoff from pasture. Soil Biology and Biochemistry 11:459-462.

Shaw, C., and S. Pawluk. 1986. Faecal microbiology of Octolasion tyrtaeum, Aporrectodia turgida and Lumbricus terrestris and its relation to the carbon budgets of three artificial soils. Pedobiology 29:377-389.

Shipitalo, M. J., and W. M. Edwards. 1987. Effect of crop management on the number and size of earthworm holes preserved in a soil. Agronomy Abstracts p. 246.

Shipitalo, M. J., and R. Protz. 1988. Factors influencing the dispersibility of clay in worm casts. Soil Science Society of America Journal 52:764-769.

Shipitalo, M. J., R. Protz, and A. D. Tomlin. 1988. Effect of diet on the feeding and casting activity of Lumbricus terrestris and L. rubellus in laboratory culture. Soil Biology and Biochemistry 20:233-237.

Slater, C. S., and H. Hopp. 1947. Relation of fall protection to earthworm populations and soil physical conditions. Soil Science Society of America Proceedings 12:508-511.

Spiers, G. A., D. Gagnon, G. E. Nason, E. C. Packee, and J. D. Lousier. 1986. Effects and importance of indigenous earthworms on decomposition and nutrient cycling in coastal forest ecosystems. Canadian Journal of Forest Research 16:983-990.

Springett, J. A. 1983. Effect of five species of earthworm on some soil properties. Journal of Applied Ecology 20:865-872.

Springett, J. A. 1985. Effect of introducing Allolobophora longa Ude on root distribution and some soil properties in New Zealand pastures. In Ecological interactions in soil: Plants, microbes and animals, ed. A. H. Fitter, D. Atkinson, D. J. Read, and M. B. Usher, 399-405. Oxford, England: Blackwell Scientific Publishers.

Springett, J. A., and J. K. Syers. 1979. The effect of earthworm casts on ryegrass seedlings. In Proceedings of the second Australasian conference on grassland invertebrate ecology, ed. T. K. Crosby and R. P. Pottinger, 44-47. Wellington, New Zealand: Government Printer.

Standen, V. 1984. Production and diversity of enchytraeids, earthworms, and plants in fertilized hay meadow plots. Journal of Applied Ecology 21:293-312.

Stockdill, S. M. J. 1982. Effects of introduced earthworms on the productivity of New Zealand pastures. Pedobiology 24:29-35.

Syers, J. K., A. N. Sharpley, and D. R. Keeney. 1979. Cycling of nitrogen by surface-casting earthworms in a pasture ecosystem. Soil Biology and Biochemistry 11:181-185.

Timmenga, H. J. 1987. The transport of mineral and organic matter into the soil profile by Lumbricus rubellus Hoffmeister. Ph.D. dissertation, University of British Columbia, Vancouver.

Uvarov, A. 1982. Decomposition of clover green matter in an arable soil in the Moscow region. Pedobiology 24:9-21.

van Rhee, J. A. 1963. Earthworm activities and the breakdown of organic matter in agricultural soils. In Soil organisms, ed. J. Doeksen and J. van der Drift, 55-59. Amsterdam: North Holland Publishers.

van Rhee, J. A. 1965. Earthworm activity and plant growth in artificial cultures. Plant and Soil 22:45-48.

van Rhee, J. A. 1977. A study of the effect of earthworms on orchard productivity. Pedobiology 17:107-114.

Vimmerstedt, J. P., and T. H. Finney. 1973. Impact of earthworm introduction on litter burial and nutrient distribution in Ohio strip-mine spoil banks. Soil Science Society of America Proceedings 37:388-391.

Wang, Juang, J. S. Hesketh, and J. T. Wooley. 1986. Preexisting channels and soybean rooting patterns. Soil Science 141:432-437.

Zachmann, J. E., D. R. Linden, and C. E. Clapp. 1987. Macroporous infiltration and redistribution as affected by earthworms, tillage and residue. Soil Science Society of America Journal 51:1580-1586.

9   Shoot/Root Relationships and Bioregulation

Plants and roots are of many sizes and shapes, and they develop within a wide range of environments. Regardless of their various sizes and shapes, roots generally serve to anchor plants and to absorb water and nutrients. Some store food reserves that survive the winter and support early spring shoot growth.

When considering shoot/root relationships, one should realize that plants have evolved over many years and that each plant is genetically programmed for a number of alternative developmental patterns. That is, various genes will be activated or repressed by environmental factors such as day length, nitrogen availability, and light spectral shifts associated with plant population density. Also, the strategy of the individual plant must be to survive long enough to reproduce the next generation. Therefore, the plant must be able to detect and adapt to various environmental situations and to partition enough photoassimilate to the roots to support shoot growth and development under those conditions.

It is apparent that plants have evolved to "invest" photoassimilate where it will best contribute to survival of the plant and its reproduction in a given environment. For example, it has been observed that genetically identical plants grow quite differently in dense populations than in sparse populations, in fertile soil than in infertile soil, or in spring than in autumn.

As a plant physiologist, I ask how much root is needed, and how does the plant sense environmental variables that regulate partitioning of photoassimilate between shoots and roots? Also, is a more extensive root system always better, and how can we use this information in field crop management?

217

As with most biological responses, it appears that a combination of genetic and environmental factors serve as natural regulators. The remainder of this chapter will be devoted to a discussion of the regulation of shoot/root relationships.

GENETIC CONTROL

There are many differences in root development among plant species. Some of the most obvious are, of course, the fibrous roots of forage grasses contrasted with the roots of plants such as soybean (<u>Glycine</u> <u>max</u>) and cotton (<u>Gossypium</u> <u>hirsutum</u>). Within a species, some characteristics may serve as survival mechanisms under specific conditions while going undetected under other conditions. For example, the sandy soils of the southeastern coastal plain of the United States often overlay a hardpan that blocks penetration of most roots and thereby limits the rooting zone. As part of our research toward improved crop efficiency on such soil, we identified some cotton genotypes that grew well, while others wilted severely when grown over a subsoil hardpan in field plots without irrigation during the drought of 1986. In subsequent controlled-environment studies, roots of a genotype that grew well in the field test penetrated an artificially compacted soil layer, while a genotype that wilted in the field test failed to penetrate the compacted soil layer. This is an example of a genotypic difference in rooting characteristics that can express under specific conditions (the compacted subsoil layer in this example) while no rooting differences are apparent between the two genotypes in the absence of the compacted layer. This type of information should be useful in developing varieties for cropping systems that are less dependent on irrigation.

In another example of genotypic differences in rooting within a species, we have tissue culture regenerated tall fescue (<u>Festuca</u> <u>arundinacea</u>) plants with root characteristics that range from very fine to very coarse, even though the shoot growth appears to be the same. Some of these genetic lines (somaclonal variants) may prove to be superior under specific soil and water conditions. Culture and identification of superior somaclonal variants for a specific purpose is a possible agricultural benefit of plant biotechnology.

ENVIRONMENTALLY INDUCED REGULATION

As plants evolved, they developed capability to detect various factors of the environment and to regulate growth processes to favor survival in that environment long enough to reproduce the next generation. Some of the dominant environmental factors that regulate morphological development include nutrient and water availability, day length, and the spectral distribution of light associated with, for example, competition from other growing plants.

Nitrogen Availability

The influence of adequate or inadequate nitrogen is shown in Figure 9.1. The two sunflower plants were started from seeds that germinated in nutrient-free sand on the same date. The seedlings were transferred to hydroponic nutrient cultures when cotyledons opened. The plant on the left grew in an aerated "complete" nutrient solution, while the one on the right grew under identical conditions except that the nitrogen was withheld from the nutrient solution. The plant on the left was obviously healthy even though it had a small root system. With no water or nutrient limitations (and the same light environment), the plant invested most of its new photosynthate in shoot growth. From a survival standpoint, this growth strategy did not "waste" excess photoassimilate on an unnecessarily large root system. Instead, investment of more of the photoassimilate in larger leaves and stem increased the photosynthetic area, which led to a larger plant that could produce more seed. The nitrogen deficiency affected partitioning toward roots. It appears to be a survival response triggered by a stress factor. This rather simple experiment demonstrates adaptation of a plant to favor survival under a specific set of conditions.

From a practical standpoint, the plant grown on the complete nutrient solution demonstrates what could happen in a field situation when water and nutrients are metered into the root zone via a trickle irrigation system. This prioritization of photoassimilate partitioning to the shoots would appear to be desirable, unless (1) the top-heavy, poorly rooted plants were blown over by a high-velocity wind, or (2) the constant flow of water and nutrients was abruptly interrupted, which could cause severe damage before the plants could adjust to the stress condi-

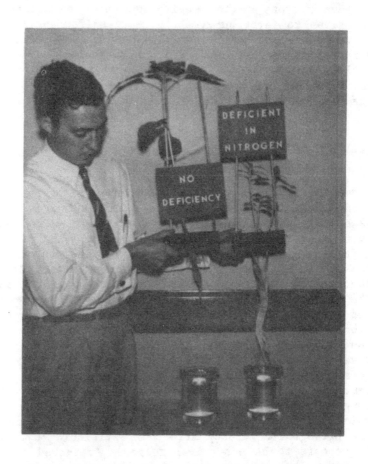

Figure 9.1.   Influence of nitrogen deficiency on seedling shoot/root relationships.

tions.   On the other hand, a plant that began growth under somewhat stressed conditions could readily adapt to adequate, but not excessive, soil water and nutrients.   This is basically what is involved in the practice of "hardening" seedlings before transplanting them from a protected nursery to a field, where the plants adapt to the new set of environmental variables.

Photoperiod

Many plants are able to "measure" the day length (actually the period of uninterrupted darkness) to trigger the induction of flowering so that seeds can develop and ripen before freezing weather occurs. Many short-day annual plants such as cocklebur (Xanthium pensylvanicum) may germinate and start growth at various times during the season, and still flower at the same time. They are photoperiodically sensitive, and even though plants are of different ages and sizes when the "critical" photoperiod occurs, a sensing system within the plant causes the flowering process to begin. The plant that started growth in early spring would usually grow larger and be capable of supporting more flowers and seed. From a survival standpoint, however, both early and late plants would produce some seed for the next generation.

Biennial long-day plants such as sweetclover (Melilotus alba) also are able to measure photoperiod and regulate partitioning of photoassimilate to favor survival. During the seedling year, the seeds usually germinate in spring and develop shoots with relatively small tap roots while days are reasonably long. As days become shorter in late summer and early autumn, the shoots seem to stop growing while the tap roots enlarge rapidly and develop crown buds. The shoots may freeze in winter. New shoots develop rapidly from the crown buds during the following spring, at the expense of the stored reserves in the large tap roots. Figure 9.2 shows the rapid development of biennial sweetclover tap roots grown in a field at Ames, Iowa, and collected at monthly intervals beginning in mid-August. A rapid change in shoot/root ratio occurred during this period of naturally decreasing photoperiods and temperatures. In a parallel experiment to compare regulatory effects of naturally decreasing photoperiods under warm conditions, some sweetclover plants were moved intact (in blocks of soil) from the field to the soil bed of a greenhouse in mid-August. Both field and greenhouse received naturally decreasing photoperiods. However, the greenhouse minimum temperature was 22°C, while field temperature approached freezing in late October and November. As shown in Figure

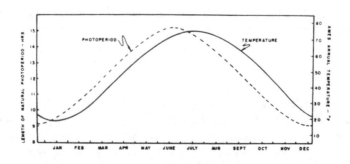

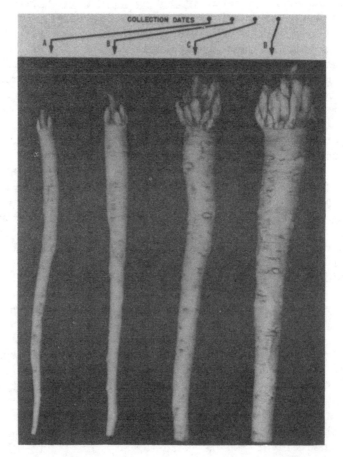

Figure 9.2. Field-grown biennial sweetclover tap roots dug August 20, September 20, October 20, and November 20.

Source: Kasperbauer 1962.

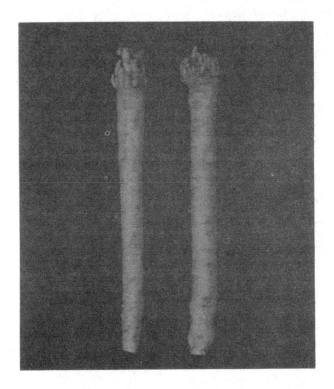

Figure 9.3. Tap roots from first year biennial sweetclover plants grown on natural photoperiods with natural (left) and 22°C minimum (right) temperatures until November.

Source: Kasperbauer 1963.

9.3, tap roots from both field and greenhouse sites were about the same size in mid-November. The tap root enlargement was dominated by photoperiodic control. The plants shown in Figure 9.4 were grown on four different photoperiods in a warm greenhouse. All were started from the same lot of biennial seed on the same day.

Those grown under the longest photoperiods flowered early and did not develop enlarged tap roots. In contrast, the plants grown on nine-hour photoperiods developed only low-growing shoots and large, fleshy tap roots. The appar-

Figure 9.4.   Plant size and flowering condition (top) and
root size (bottom) of biennial sweetclover plants after 100
days (from germination) of exposure to photoperiod treatment.   Left to right: 24-hour, 20-hour, 16-hour, and 9-hour
photoperiods in a warm greenhouse, grown in 10-cm clay
pots.

Source: Kasperbauer 1963.

ent "signal" from the short photoperiod was that winter was coming and there would not be enough growing days to allow the plants to flower and develop ripe seed before a killing frost. In contrast, those on the longest days received a photoperiodic "signal" that there was plenty of time to flower and develop ripe seed before winter. Consequently, they did not form storage roots because the life cycle (from seed to the next generation of seed) was completed.

Knowledge of photoperiodic regulation of shoot/root relationships is important in management of root crops. It is especially relevant in the use of biennial legumes as "green manure" crops to incorporate organic matter for soil improvement. An important point is that even though the shoots of biennial legume plants seem to stop growth late in the season, root enlargement continues with the decreasing day lengths, until the shoots freeze.

Light Spectral Distribution

Spectral composition of light can alter the shoot/root ratios of developing plants. The regulatory mechanism and plant responses have been studied in detail under controlled environments. Recent research has shown that the same light-sensing mechanism also responds to naturally occurring spectral differences associated with plant population density (nearness of competing plants), row orientation (especially in broadleaf plants), and soil or mulch color under field conditions. It is now apparent that plants are capable of sensing competition from nearby plants and to modify developmental patterns according to the amount of competition (Kasperbauer 1987, 1988).

Controlled-environment studies have shown that the ratio of light received at 735 nm (called far-red and usually designated FR) relative to that received at 645 nm (called red and designated R) is measured by a photoreversible pigment (phytochrome) within the growing plant. Phytochrome is present in minute quantities relative to chlorophyll and the carotenoids, and the greatest concentrations of phytochrome are present in regions of actively dividing or recently divided cells. While light from 400 to 700 nm is absorbed by photosynthetic pigments and results in production of photosynthate, the photoequilibrium level of phytochrome (as regulated by the FR/R ratio) appears to play a major role in partitioning and use of photosynthate within the plant, as an adaptation to environmental conditions.

In a controlled-environment experiment summarized in Table 9.1, a high FR/R ratio caused soybean seedlings to develop longer stems and smaller root systems. Parallel experiments with wheat (<u>Triticum</u> <u>aestivum</u>) and tobacco (<u>Nicotiana</u> <u>tabacum</u>) resulted in the same trends. That is, a higher FR/R ratio resulted in longer stems, fewer lateral branches or tillers, a smaller root system, and a higher shoot/root dry matter ratio. If such a system could function under field conditions to sense competition from other plants and regulate partitioning of photoassimilate among plant components, it might serve as a regulator of plant adaptation to competition and favor survival.

Under field conditions, we measured the spectral distribution of incoming sunlight and compared it with spectral distribution of light received at various points in plant canopies grown in different population densities and row orientations. As expected, the various wavelengths of light were not absorbed equally by growing plants. Examination of a typical soybean leaf showed that the leaf absorbed most of the visible light and reflected or transmitted most of the FR (Figure 9.5). The same patterns were found for tobacco, tomato (<u>Licopersicon</u> <u>esculentum</u>), corn (<u>Zea</u> <u>mays</u>), and wheat, supporting the concept that a common regulatory mechanism might exist among these species. Since each green leaf reflected FR, it was reasonable to

TABLE 9.1.   Influence of FR/R ratio on soybean shoot/root relationships.

FR/R ratio* at end of photosynthetic period	Dry matter distribution			
	Shoots		Roots	
	Leaf blades	Stems + petioles	Roots	Nodules
	------------------(%)------------------			
Low	43.9	23.6	30.1	2.4
High	43.6	33.2	21.3	1.9

*All seedlings were grown in the same controlled-environment chamber. The only treatment difference was that half of the plants received a high and the other half received a low FR/R light ratio for five minutes at the end of the daily photosynthetic period for twenty consecutive days.

<u>Source:</u> Kasperbauer 1987

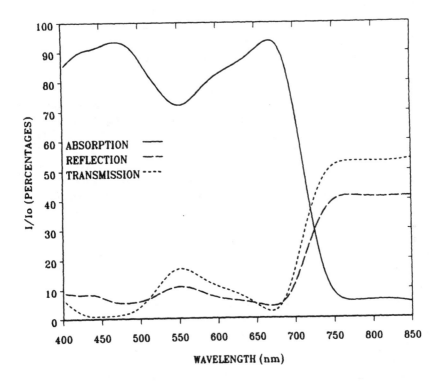

Figure 9.5. Absorption, transmission, and reflection of light from a typical soybean leaf. I/Io refers to radiation absorbed, transmitted, and reflected at five-nanometer intervals relative to incident radiation at the same wavelengths.

Source: Kasperbauer 1987

expect that a plant surrounded by many other plants (FR reflectors) would receive a higher FR/R ratio than an isolated plant or one in a low population density.

Subsequent light measurements in plant population density, row-orientation, crop species, and plant age studies clearly showed that a plant growing in a high plant population density received a higher FR/R ratio than one growing in a low population density. Also, broadleaf plants, such

as soybean and bush bean (<u>Phaseolus</u> <u>vulgaris</u>), growing in
north-south as compared with east-west rows received
slightly higher FR/R ratios because of heliotropic movement
of the leaves (that is, individual leaves became direction-
al FR reflectors). The row-orientation effects on FR/R ratio in corn plots were less pronounced. As predicted from
the controlled-environment studies, plants that received
higher FR/R ratios in the field developed longer stems, fewer lateral branches or tillers, longer and narrower leaves,
and smaller root systems. That is, the FR/R and shoot/root
ratios were highly influenced by the nearness of competing
plants. This pattern occurred even when the roots developed in root-tight containers (embedded in the soil) such
that plants in all field plant spacings and row orientations had the same volume of rooting medium. The indicated
pattern of morphological development in the field was already evident in seedlings soon after emergence, long before mutual shading and reduced photosynthetically active
light became a factor. Very young seedlings are highly sensitive to the FR/R ratio, and they are dramatically influenced by even subtle changes in the ratio. The sensing of
competition from other plants and the partitioning of photoassimilate to the stem (at the expense of lateral branches
or roots) in a high plant population density would allow a
plant to increase its probability of keeping some leaves in
sunlight above competing plants and of surviving and producing some seed. On the other hand, a low FR/R ratio (as
would occur in a low population density) would favor increased partitioning to branches and roots. This adaptation to low population density should allow the individual
plant to support development of more seed.

In the foregoing examples, growing plants responded to
the FR/R ratio supplied by lamps and filters in controlled
environments and to amounts of reflected FR associated with
the nearness and number of competing plants in field studies. It is apparent that plants have evolved to respond
differently to various wavelength combinations, and it is
also apparent that the plants cannot discern the source of
the spectral alterations. This line of reasoning led us to
consider developmental effects of the spectral distribution
of upwardly reflected light from variously colored soils
and mulches. We found that plants developed different
shoot/root ratios when grown over different soil surface
colors, even when the root temperatures were kept constant
by use of insulation panels below the various soil surface
colors (Hunt et al. 1989). It is evident that the spectral
distribution of reflected light can influence photoassimi-

late partitioning within a plant and affect its shoot/root relationships.

The use of variously colored mulches and plant residue covers to modify the reflected light spectrum and the partitioning of photoassimilate among plant components (without interfering with incoming sunlight) appears to offer opportunity for increasing the quantity and quality of crop productivity. One mulch color may favor leaf crops, another may favor fruit, and another may favor root crops. Further knowledge of the bioregulatory role of light and its manipulation under field conditions appears to have great potential for future crop production systems.

In summary, the growth and development of a plant is the result of its genetics and the environment within which it is grown. Plants have evolved the capability to sense various environmental factors and to activate or repress genes to regulate developmental patterns that favor survival of the plant long enough to produce the next generation. The relationship between the quantity of visible light and the amount of photosynthesis has been widely studied. However, awareness of the role of light reflected from competing plants or from different colors of soil or mulch to regulate partitioning of photoassimilate is just now being realized. My question concerning the best root size for a plant is still unanswered. Certainly, the largest root system is not always on the largest plant. Better understanding of the natural bioregulation of shoot/root relationships under field conditions will be highly useful in future plant-soil-water-light management systems.

ACKNOWLEDGMENTS

Examples discussed in the chapter are chiefly from my collaborative work with H. A. Borthwick, F. P. Gardner, S. B. Hendricks, P. G. Hunt, I. J. Johnson, D. L. Karlen, K. Kaul, W. E. Loomis, T. A. Matheny, and W. Sanders.

REFERENCES

Decoteau, D. R., M. J. Kasperbauer, and P. G. Hunt. 1989. Mulch surface color effects on yield of fresh-market tomatoes. Journal American Society of Horticultural Science 114:216-219.

Hunt, P. G., M. J. Kasperbauer, and T. A. Matheny. 1987. Nodule development in a split-root system in response to red and far-red light treatment of soybean shoots. Crop Science 27:973-976.

Hunt, P. G., M. J. Kasperbauer, and T. A. Matheny. 1989. Soybean seedling growth responses to light reflected from different colored soils. Crop Science 29:130-133.

Kasperbauer, M. J. 1986. Tall fescue plant modification through tissue culture, haploids and doubled haploids. In Proceedings of sixth international congress of plant tissue and cell culture, ed. D. A. Somers, B. G. Gengenbach, D. D. Beisboer, W. P. Hackett, and C. E. Green, 124. Minneapolis: University of Minnesota Press.

Kasperbauer, M. J. 1987. Far-red reflection from green leaves and effects on phytochrome-mediated assimilate partitioning under field conditions. Plant Physiology 85:350-354.

Kasperbauer, M. J. 1988. Phytochrome involvement in regulation of photosynthetic apparatus and plant adaptation. Plant Physiology Biochemistry 26(4):519-524.

Kasperbauer, M. J., H. A. Borthwick, and S. B. Hendricks. 1964. Reversion of phytochrome 730 (Pfr) to P660 (Pr) in Chenopodium rubrum L. Botanical Gazette 125(2):75-80.

Kasperbauer, M. J., R. C. Buckner, and W. D. Springer. 1980. Haploid plants by anther-panicle culture of tall fescue. Crop Science 20:103-106.

Kasperbauer, M. J., and G. C. Eizenga. 1985. Tall fescue doubled haploids via tissue culture and plant regeneration. Crop Science 25:1091-1095.

Kasperbauer, M. J., and F. P. Gardner. 1962. Day length controls root size and flowering in sweetclover plants. Crops and Soils June 62:21.

Kasperbauer, M. J., F. P. Gardner, and I. J. Johnson. 1963. Taproot growth and crown bud development in biennial sweetclover as related to photoperiod and temperature. Crop Science 3:4-7.

Kasperbauer, M. J., F. P. Gardner, and W. E. Loomis. 1962. Interaction of photoperiod and vernalization in flower-

ing of sweetclover (<u>Melilotus</u>). <u>Plant Physiology</u>
37:165-170.

Kasperbauer, M. J., and P. G. Hunt. 1987. Soil color and
surface residue effects on seedling light environment.
<u>Plant and Soil</u> 97:295-298.

Kasperbauer, M. J., and D. L. Karlen. 1986. Light-mediated
bioregulation of tillering and photosynthate partitioning in wheat. <u>Physiologia Plantarum</u> 66:159-163.

Kaul, K., and M. J. Kasperbauer. 1988. Row orientation
effects on FR/R light ratio, growth and development of
field-grown bush bean. <u>Physiologia Plantarum</u>
74:415-417.

B. L. McMichael

10 Root-Shoot Relationships in Cotton

The distribution of assimilates between root and shoots of the cotton plant is a complex process. It can have a profound influence on the growth and productivity of the plant. This distribution is influenced by above- and below-ground factors that may be environmental in nature or mediated by cultural practices. If, for example, there is a decrease in partitioning of photosynthate to the developing fruit because of the failure of the plant to develop an adequate root system in time to alleviate soil water deficits, then productivity may be severely reduced. Thus, the coordination between the growth of the shoots and the roots is necessary to maintain productivity, since the development of either part is dependent on the other. The study of the interactions between root and shoot growth in cotton has been very difficult to do. In many studies, where shoot growth is documented, adequate information regarding growth and development of the root systems is generally lacking. The information that is available provides some insight into the complex nature of biomass distribution that will be discussed in the following sections. Since the measurement of the biomass present in the root system is critical to the evaluation of root-shoot relationships, a discussion follows of current measurement techniques and of some factors that influence root-shoot relations in cotton. Further information will be presented on the dynamic changes in root-shoot ratios and the potential for using observed genetic diversity for improvement of root-shoot relationships.

SHOOT DEVELOPMENT

Several publications (Tharp 1965; Mauney 1986;
Constable and Gleeson 1977) describe the development of
shoots of the cotton plant. The stem develops during
growth of successive nodes above the cotyledonary nodes.
Production of nodes on the main stem (sympodial node)
occurs with a somewhat regular frequency (one node every
2-4 days under most conditions). The total number of
main-stem nodes a plant will produce during a growing
season will vary according to various environmental
conditions such as soil water availability and temperature
(Mauney 1986). It is not uncommon for 25-30 nodes to
develop during a typical growing season. A true or
main-stem leaf initiates at each main-stem node with two,
or possibly three, axillary buds. The first axillary bud
can produce either a vegetative or fruiting branch. If
something should happen to abort the first bud the other
axillary bud or buds may take over and produce a branch
(Tharp 1965; Mauney 1986).
Vegetative (monopodial) branches are produced near the
base of the plant and up to the fifth or sixth main-stem
node; after which fruiting branches are produced on the
successive nodes. The vegetative branches grow much like
the main stem, with each branch having a separate growing
point at the terminal of the branch. The fruiting branches
have a different growth habit (Tharp 1965). The growing
point of each fruiting branch ends at each flower produced
on the branch. The branch continues to grow and thus
develops more flowers from the axillary bud formed with
each flower. Therefore, the fruiting branches grow in a
zig-zag pattern as each section of the branch develops from
each successive axillary bud at the end of the branch. A
subtending leaf is also produced next to each flower and is
important in providing photosynthate to the developing
fruit or boll (Wullschleger and Oosterhuis, in press). The
first flower on the first fruiting branch will generally
open 45-50 days after seedling emergence, depending on a
variety of growing conditions. When this occurs, the first
flower on the next highest fruiting branch will generally
open three days later (Tharp 1965; Mauney 1986). Flowers
on the same fruiting branch will open about six days apart.
Once the flower has opened and has been pollinated, the
developing boll usually requires 50-60 days to mature.
Therefore, there are several bolls and flowers present on

the plant at the same time that are in various stages of growth and maturity.

Development rate of the shoots, in terms of dry matter accumulation of shoot components, is shown in Figure 10.1 (Jordan 1982). The temporal relationship of plant height, leaf area, fruit weight, and root length are shown in Figure 10.2 (McMichael 1980). In general, the rate of leaf and stem development was reduced as the bolls begin to grow and mature. These developing bolls are a large sink for the photosynthate produced in the leaves. The total biomass produced by the shoots, including the bolls, accounts for about 80-90 percent of the total biomass produced by the plant (De Souza 1987; Dilbeck, Quisenberry, and McMichael 1984).

DEVELOPMENT OF THE ROOT SYSTEM

Development of the cotton plant root system has been described in several research papers and review articles (McMichael and Quisenberry 1986; Hayward 1938; Tharp 1965; Pearson 1974; Taylor and Klepper 1978). The root system is comprised of a primary root or taproot and many branch or lateral roots. First-order lateral roots initiating from the taproot give rise to higher order lateral roots. Initially, the taproot may grow several days after germination without branching. When branching does begin, usually four to five days after germination, the lateral root primordia develop about 10-12 cm behind the tip of the taproot (McMichael 1986; Mauney 1968). Depth of taproot penetration may vary from less than 1 meter to as much as 3 meters, depending on many factors such as the degree of soil compaction and genetic characteristics of the variety (Taylor and Gardner 1963; Quisenberry et al. 1981). Therefore, the rate of elongation of the taproot may range from less than 1 cm per day to as much as 6 cm per day (Taylor and Ratliff 1969).

The lateral roots usually remain somewhat shallow in the soil (less than 1m). It has been shown that the cotton plants lateral roots are predominately horizontally oriented (Upchurch, McMichael, and Taylor 1988). The extensibility of the root system depends a great deal on the initiation and growth of higher order lateral roots. Therefore, the more fibrous the root system can become, the greater is the potential for a higher rooting density and a large soil volume occupied by roots.

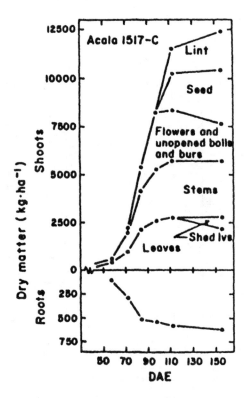

Figure 10.1. Distribution of dry matter in 'Acala 1517-C' cotton as a function of days after emergence.

<u>Source</u>: W. R. Jordan 1983 (from Halevy, 1976).

Development of the cotton root system, in terms of dry matter production, is depicted in Figure 10.1 and in Figure 10.3. Data for Figure 10.1 were collected from a field study. The root biomass was measured by digging around the plant and collecting most of the large roots and a lesser portion of the small roots. The root biomass shown in Figure 10.3 was measured by growing plants in large containers and washing soil away from the entire root system. In general, total biomass produced by the root system comprises approximately 10 percent of total biomass produced by the plant during a growing season (Jordan 1982; Dilbecket al. 1984). Total root length produced during the same period may be several hundred meters (McMichael 1988; Taylor and Klepper 1974). In the developmental process, both total length and presumably dry weight increases as the plant grows, and until the maximum height is reached and the bolls begin to develop (Figure 10.2) (McMichael

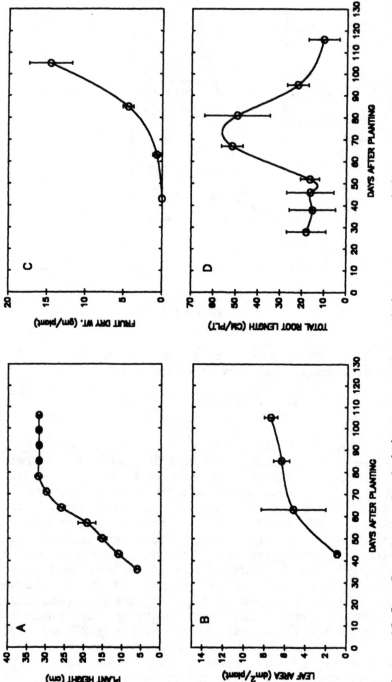

Figure 10.2. Development of (A) plant height, (B) leaf area, (C) fruit weight, and (D) root length of field grown 'Acala SJ-5' cotton grown under dryland conditions in 1980. <u>Source</u>: B. L. McMichael, unpublished data.

1980; Pearson and Lund 1968). Net root length then begins to decline as older roots die, while total root biomass may remain somewhat constant.

MEASUREMENT OF ROOT-SHOOT RATIOS IN COTTON

The major problem of obtaining root-shoot ratios of plants is the collection of the total root biomass. Generally, studies are conducted in a greenhouse or a growth chamber, depending on the particular experiment, where plants are grown in containers of soil or in hydroponics (solution culture). If plants are grown in soil, or another potting medium, roots must be washed free of the medium before their biomass can be measured. There are several techniques described in the literature for washing out root systems that provide for minimum loss of the root material (Böhm 1979; Brown and Thilenius 1976; Smucker et al. 1982). Most of these methods involve placing the soil-root mass onto a fine mesh screen and washing the soil through the screen. Roots retained on the screen are collected for drying and weighing. Some equipment is also commercially available for separating roots from soil, especially for small samples or soil cores.

If plants are grown in solution culture, it is a simple matter to collect the roots for biomass determination, since no separation of the roots from the growing medium is necessary. Some researchers contend that roots grown in solution culture do not mimic soil-grown roots, both in terms of rates of biomass production and in development of total length. The author knows of no studies that directly compares biomass production of soil-grown versus hydroponically grown roots. A speculation would be that, for plants grown under similar conditions, root-shoot ratios of the soil-grown plants may be higher than the hydroponically grown plants. This may be due to stresses generated in the soil-grown plants because of the pressure required to grow roots through the soil matrix.

There have been very few field studies involving cotton measurements that partition biomass into roots and shoots (Halevy 1976). There have been many studies that describe development of the aerial portion of plants in terms of total biomass production, and that partition biomass into leaves, stems, and fruit (Constable and Gleeson 1977; Hearn 1969; Bruce and Romkens 1965; Crowther 1941). There was no attempt in these studies to measure

238

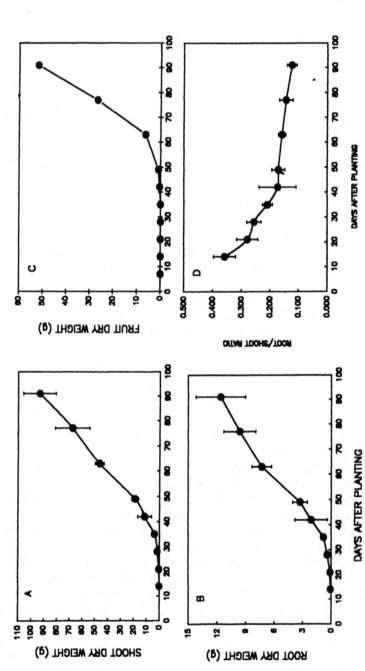

Figure 10.3. Development of (A) shoot weight, (B) root weight, (C) fruit weight, and (D) root-shoot ratios in greenhouse-grown "Paymaster 145" cotton. Plants were well-watered throughout the experiments.

Source: B.L. McMichael, unpublished data.

biomass production of roots, or even attempt an estimate of total root biomass. Therefore, the following discussions of the dynamics of root-shoot relationships and the influence of various environmental factors on partitioning of biomass in cotton will be confined to studies conducted in situations (greenhouse and controlled-environment chambers) where root biomass production was evaluated.

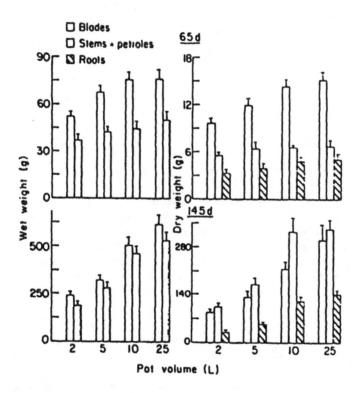

Figure 10.4. Partitioning of fresh and dry weights of organs of 'Acala SJ-2' cotton, on a per/plant basis, grown in pots of different capacities. Plants were sampled at 65 and 145 days after planting. One SE is shown at the top of each bar.

Source: A. Carmi and J. Shalhevet 1983. Reprinted by permission.

DYNAMICS OF ROOT-SHOOT RELATIONSHIPS

Davidson (1969) describes dynamics of root-shoot relationships by the equation: Root mass x root activity = Shoot mass x shoot activity. He defines root activity as water and nutrient absorption and shoot activity as photosynthesis. Thus, there is a balance between roots and shoots that usually is maintained, because of some stimuli such as fruit abscission or root pruning, by selective partitioning of carbohydrates.

In many earlier studies, where biomass of both roots and shoots were measured, the root biomass was measured by destructive harvesting at the end of the experiment. For example, Eaton (1955) describes experiments where cotton plants were grown in large containers and harvested six months later. His results showed that root biomass made up 12-14 percent of total biomass, and that root-shoot ratios were about 0.22. Similar procedures were used by Eaton and Joham (1944) in their studies on the effects of defruiting cotton plants on root and shoot development. Other studies, (Carmi and Shalhevet 1983; De Souza and Da Silva 1987), reported information on root biomass production at least twice during their experiments (Figure 10.4 and Table 10.1). McMichael (1988) grew plants in large containers in the glasshouse, and harvests were made either weekly or bi-weekly from 14 days after planting to 90 days after planting. The results, representing one of the few studies where the dynamics of root-shoot relationships were measured over a long period, are summarized in Figure 10.3. The increase in root biomass (Figure 10.3B) was linear from 30-40 days after planting up to approximately 60-70 days after planting. After that the rate of increase was reduced as significant increases in fruit development (Figure 10.3C) occurred. Fruit development constituted a large sink for photosynthate. Root-shoot ratios decreased rapidly during the early stages of plant development (Figure 10.3D), suggesting a more rapid development of shoot biomass during that time (Figure 10.3A). The values for root-shoot ratios then remained somewhat constant, or decreased slightly, for the remainder of the growth period (day 50 to day 90). These results are consistent with the results reported earlier by others (Halevy 1976; Jordan 1982; Eaton 1931.)

TABLE 10.1. Dry Weight, Starch Concentrations, and Root-Shoot Relationships Between an Annual and Perennial Variety of Cotton and at two Harvest Dates.

	42 d.		80 d.	
	Annual	Perennial	Annual	Perennial
---------------------------- g ----------------------------				
Shoot dry weight	10.22a	5.94	43.73a	31.75b
Root dry weight	2.61a	1.81b	7.75b	15.08a
Root/shoot ratio	0.261b	0.306a	0.177b	0.491a
----------------- mg g^{-1} dry matter -----------------				
Starch				
Tap root	64.06b	106.12a	138.90b	196.97a
Lateral roots	21.05b	27.38a	31.83b	67.00a

Source: J. G. De Souza and J. V. Da Silvia 1987.

They were also consistent with the results for development of root length over a growing season (McMichael 1980, Figure 10.2D). It should be noted that many studies have reported changes in root length, or root length density (cm root cm^{-3} soil), with time relative to development of above-ground portions of the plant (Kennedy et al. 1987; Klepper et al. 1973; Grimes 1975; Taylor and Klepper 1974). But, there have not been, to the author's knowledge, any definitive studies conducted that directly relate the development of root length to the development of root biomass. This is presumably because that, at any given time, the root system of cotton is comprised of many different size roots. They may range from less than one millimeter in diameter to several millimeters, causing the weight per unit length to change with time. This restriction is evident in the discussions of the effects of different environments and various treatments on root-shoot relationships.

FACTORS INFLUENCING ROOT-SHOOT RELATIONSHIPS IN COTTON

There are many factors that influence the interactive growth patterns of cotton shoots and roots, and their interactive growth habits. A few of the major factors will be discussed briefly. The reader is encouraged to consult reviews by Jordan (1982) and Mauney and Stewart (1986) for a more in-depth discussion of cotton plant growth and references relating environmental influences on shoot and root growth.

Temperature

Temperature of the soil and the air surrounding the above-ground portion of the plant has a significant influence on the growth of cotton shoots and roots. Hutmacher et al. (1986) observed that as the air temperature increased, beginning at 20-22°C, the gross photosynthetic rates of cotton leaves increased up to 30-32°C. An increase in temperature above that point caused the rates of photosynthesis to decrease. He also observed in the same experiments that net photosynthesis (net carbon exchange rate) decreased as the air temperature increased, and that photorespiration increased dramatically. Mauney (1966) and Hesketh et al. (1972a) showed that shoot growth and main-stem node production was affected by temperature. They observed that the node production rate was significantly lower at low temperatures (18-20°C), which could ultimately impact shoot biomass production. Wanjura and Buxton (1972) found that elongation of both the hypocotyl and radicle of cotton seedlings increased as soil temperatures increased from about 15 to 32°C. They observed a decrease in elongation rates when the temperature reached 37-38°C. Fruit production in cotton is temperature dependent. Mauney and Phillips (1963) noted that both day and night temperatures had an effect on the site of the first fruiting node. Hesketh et al. (1972a) found that the rate of maturation of squares (preflowering fruit forms) was three times as high at 39°C as it was at 18°C. Gipson and Joham (1968a,b) found that both flowering and fruit retention increased at low night temperatures and that boll development was adversely affected. Mauney (1986), Christiansen et al. (1986), and Gipson and Shaw (1986) provided excellent reviews describing the influence of temperature on fruit development and fiber production.

The results they report are important to our discussion of root-shoot relationships in cotton.

Several studies have indicated that the optimum temperature range for root growth in cotton is somewhere between 28°C and 35°C (Taylor 1972; Lety et al. 1961). When soil temperature dropped below this range, root growth was reduced and less branching occurred (Brouwer and Hoagland 1964; McMichael 1988). The shoot-root ratio increased when temperature in the root zone was lowered (Nelson 1967). If soil temperature was much higher than 35°C metabolic activity and elongation rates were reduced (Neilson and Cunningham 1964; Taylor 1983).

Soil Strength

Soil physical properties may have a significant influence on cotton root-shoot relationships. Generally, increases in soil strength result in a decreased root growth, while shoot growth may be reduced or remain the same depending on other environmental factors. Wanjura and Buxton (1972) reported that an increase in the physical impedance of soil decreased the elongation of the hypocotyls and radicles. Their results also showed that hypocotyls were more sensitive to impedance than radicles. Taylor and Ratliff (1969) showed that the elongation rate of cotton roots decreased as soil strength is increased. Ben-Porath (1985), working with cotton grown in different soil volumes, found that there was a more determinate growth pattern in the root-restricted plants. Before blooming the shoots grew faster than the roots. He also observed an increase in yield in the root-restricted plants due to a shift in partitioning of assimilates into fruit rather than into roots or other plant organs. Carmi and Shalhevet (1983) reported similar results with cotton plants growing in different size containers placed in the field. Taylor and Gardner (1963) reported a significant negative correlation between soil strength and taproot elongation. The critical impedance factor controlling root penetration was soil strength and not soil bulk density.

Soil Water

Soil water content affects the growth and productivity of plant shoots. It has been well documented that soil water deficits may result in increased leaf and fruit abscission (McMichael, Jordan, and Powell 1973; McMichael and Elmore, 1976), depending on both the severity and the duration of the stress. Ritchie and Burnett (1971) showed that development of total shoot biomass decreased during prolonged periods of water deficits. Eaton (1955) reported similar results with cotton plants grown in large containers that were allowed to wilt periodically throughout the growth period. Therefore, the net result of a water stress is to reduce the growth of the shoots and reduce the total shoot biomass production.

Cotton root distribution may be altered because of a reduction in soil water content. Rooting depth may increase as the soil dries (Klepper et al. 1973), and elongation rates of roots may be significantly reduced (Taylor 1983). Rooting density also may decrease. Eaton (1955) noted that the total root biomass of cotton plants grown in large containers and allowed to wilt did not differ from the root biomass produced by "well-watered" plants. It is possible that the control plants were allowed to undergo a mild water stress. At any rate, the general result of a reduction in soil water content is to reduce top growth and to increase rooting depth. Thus, if root biomass increased with greater rooting depth, the root-shoot ratio would tend to increase because of the water deficit. This was the case in other species (Aung 1974).

Fruit Production

The production of fruit (bolls) in cotton has a very dramatic influence on growth of the shoots and roots. Since the initiation and growth of bolls are such a large sink for assimilates, a significant reduction in the partitioning of biomass into vegetative plant parts and roots can occur. Eaton (1931) and Eaton and Joham (1944) showed that the removal of floral buds increased the weight of the shoots by as much as 29 percent over plants that were allowed to develop normally. The mean root-shoot ratios ranged from 0.22 for normal fruiting plants to 0.53 for plants grown without bolls. The concentration of sugars in the shoot almost tripled three weeks after defruiting. The data also showed that the weight of the roots increased more than the weight of the leaves. It is clear from this information that the developing bolls have priority for current photosynthate and, without a significant fruit load, the root-shoot ratios increase due to increased root growth.

Hormone Relationships

It has been suggested that a controlling factor in root-shoot relationships, besides the source-sink factor, is the production and translocation of plant hormones. Russell (1977) pointed out that the control mechanism for partitioning dry matter into roots and shoots may be hormone mediated, since evidence for the role of plant hormones in promoting or inhibiting growth processes has been documented in several plant species in recent years. Tollervey (1970), in his review of the physiology of the cotton plant, also mentions the possibility of hormonal control of source-sink relationships in cotton that would directly affect partitioning of dry matter into roots and shoots. Although the exact mechanisms are not clear, there is evidence that auxins produced in the shoots may have an influence on the growth of the root system. It has been suggested that cytokinins produced in the roots may have an effect on the growth of the shoots (Torrey 1976; Sachs and Thimann 1967). Guinn and Brummett (1986) point out that hormones, which can significantly impact root-shoot relationships, have an important role in regulating fruit production and retention in cotton.

GENETIC ASPECTS OF ROOT-SHOOT RELATIONSHIPS

Genetic variability has been observed in both cotton shoot and root development (Bourland and White 1985; Quisenberry et al. 1981; McMichael et al. 1985; Dilbeck et al. 1984). Kennedy et al. (1987) showed that there were varietal differences in root growth into acid subsoils. Their studies showed, for example, that the Stoneville 825 and Pima S-5 cultivars grew more roots into the acidic subsoils than did Auburn 56 or Deltapine 41. They also showed that the Stoneville cultivar had a larger shoot biomass than did the Deltapine cultivar, which correlated with the root distribution. De Souza and Da Silva (1987) studied the partitioning of carbohydrate in what he designated as "annual" versus "perennial" cottons. His results show that the root-shoot ratios were highest in the perennial cottons, with preferential partitioning of dry matter into the roots. The perennial strains were of the maria galante race (<u>G. hirsutum</u>) of cotton and tend to produce an extensive root system very early in their development. In studies with several exotic cotton race stocks, Quisenberry et al. (1981) showed significant differences in shoot biomass development, taproot lengths, and number of lateral roots in 35-day-old plants. The biomass of the root systems was not determined in this study. Research by Dilbeck et al. (1984), using five male sterile A-lines and their male-fertile female counterparts (B-lines), showed that when fruit weights were included in biomass determinations there was no statistical difference between the A and B lines. The results also showed that the B-lines produced less vegetative biomass than the A-lines and that a smaller percentage of the total vegetative biomass produced by the B-lines was partitioned into stems and roots than for the A-lines. These findings showed the influence of the fruit load (B-lines) on partitioning of the total biomass. The genetic variability found in these studies suggest that the opportunity may exist for genetic alteration of root-shoot relationships in terms of biomass partitioning.

CONCLUSION

An understanding of the dynamics of carbon production and partition into roots and shoots is important to the development of more efficient water and nutrient management systems. The development of simulation models that integrate plant growth processes in relation to environmental changes have become important in recent years. They are used as a tool to aid in our understanding of factors that contribute to plant productivity. Information dealing with the varible partitioning of photosynthate into plant parts, because of changes in environmental stimuli, is vital to the continued development of more realistic growth models for evaluating potential avenues of crop improvement. Current models (Baker et al. 1983; Stapleton et al. 1973) should be updated as information dealing with partitioning into root biomass becomes available. This effort, combined with the research on the exploitation of shoot and root growth genetic variability, may eventually provide us with plants possessing under a wide range of growing conditions superior potential for productivity.

REFERENCES

Aung, L. H. 1974. Root-shoot relationships. In The plant
    root and its environment, ed. E. W. Carson. 29-63.
    Charlottesville: University of Virginia Press.
Baker, D. H., J. R. Lambert, and J. M. McKinion. 1983.
    Gossym: A simulator of cotton growth and yield. South
    Carolina Agricultural Experiment Station Bulletin no.
    1089. 134.
Ben-Porath, A. 1985. Effects of taproot restriction on
    growth and development of cotton grown under drip
    irrigation. Mississippi State University. Ph.D
    Dissertation.
Bohm, W. 1979. Methods of studying root systems. New
    York: Springer-Verlag. 188.
Bourland, F. M., and B. W. White. 1985. Variation in
    plant height and monopodial branches among selected
    cotton cultivars. In Beltwide cotton production
    research conference, 1985 proceedings, ed. T. C.
    Nelson. 85-86. Memphis, TN: National Cotton Council of
    America.
Brouwer, R. and A. Hoagland. 1964. Response to bean
    plants to root temperatures: II. Anatomical aspects.
    Jaarb. Inst. Biol. Scheik. Onderz. Landbgew Ass.
    23-31.
Brown, G. R., and J. F. Thilenius. 1976. A low-cost ma-
    chine for separation of roots from soil material. Jour-
    nal of Range Management. 29:506-507.
Bruce, R. R. and M. J. M. Romkens. 1965. Fruiting and
    growth characteristics of cotton in relations to soil
    moisture tension. Agronomy Journal. 57:135-140.
Carmi, A. and J. Shalhevet. 1983. Root effects on cotton
    growth and yield. Crop Science. 5:875-973.
Christiansen, B., A. Bationo, and J. Henao. 1986. Nitro-
    gen fertilizer efficiency in millet growing areas in
    Niger. I.F.D.C., Agronomy Abstracts. Muscle Shoals,
    Alabama.
Constable, G. A. and A. C. Gleeson. 1977. Growth and dis-
    tribution of dry matter in cotton (Gossypium hirsutum
    L.). Australian Journal of Agricultural Research.
    28:249-256.
Crowther, F. 1941. Studies in growth analysis of the cot-
    ton plant under irrigation in the Sudan. 2. Seasonal
    variation in development and yield. Annals of Botany,
    New Series. 5:509-533.
Davidson, R. L. 1969a. Effects of soil nutrients and mois-

ture on root/shoot ratios in some pasture grasses and clover. Annals of Botany, New Series. 33:531-569.

Davidson, R. L. 1969b. Effects of soil nutrients and moisture on root/shoot ratios in Lolium perenne L. and Trifolium repens L. Annals of Botany, New Series. 33:571-577.

De Souza, J. G. and J. Vieira Da Silva. 1987. Partitioning of carbohydrates in annual and perrennial cotton (Gossypium hirsutum L.), Journal of Experimental Botany. 39(192):1211-1218.

Dilbeck, R. E., J. E. Quisenberry, and B. L. McMichael. 1984. Genetic relationships between biomass production and yield in upland cotton. In Beltwide cotton production research conferences, 1984 proceedings. ed. J. M. Brown. 108-110. Memphis, TN: National Cotton Council of America.

Eaton, F. M. 1931. Root development as related to character of growth and fruitfulness of the cotton plant. Journal of Agricultural Research. 43:875-883.

Eaton, F. M. and H. E. Joham. 1944. Sugar movement to roots, mineral uptake, and the growth cycle of the cotton plant. Plant Physiology. 19:507-518.

Eaton, F. M. 1955. Physiology of the cotton plant. Annual Review Plant Physiology. 6:299-328.

Gipson, J. R. and H. E. Joham. 1968a. Influences of night temperature on growth and development of cotton (Gossypium hirsutum L.) I. Fruiting and boll development. Agronomy Journal. 60:292-295.

Gipson, J. R. and H. E. Joham. 1968b. Influences of night temperatures on growth and development of cotton (Gossypium hirsutum L.) II. Fiber properties. Agronomy Journal. 60:296-298.

Gipson, Jack, and D. Shaw. 1986. Determination of fiber properties by preharvest sampling: Texas High Plains. In Beltwide cotton production research conference, 1986 proceedings. 107.

Grimes, W. H. 1975. Industry view of pest management. In Beltwide cotton production research conference, 1975 proceedings. 127.

Guinn, G. and D. L. Brummett. 1986. ABA and IAA concentrations in young bolls and abscission zones in relation to boll retention. In Beltwide cotton production research conference, 1986 proceedings. 88.

Halevy, J. 1976. Growth rate and nutrient uptake of two cotton cultivars grown under irrigation. Agronomy Journal. 68:701-705.

Hayward, H. E. 1938. The structure of economic plants.

The Macmillan Company, New York.

Hearn, A. B. 1969. Growth and performance of cotton in a desert environment. 1. Morphological development of the crop. Journal of Agriculture, Camb. 73:65-74.

Hesketh, J. D., D. N. Baker and W. G. Duncan. 1972a. Simulation of growth and yield in cotton. II. Environmental control of morphogenesis. Crop Science. 12:436-439.

Hesketh, J. D., K. E. Fry, G. Guinn, and J. R. Mauney. 1972b. Experimental aspects of growth modeling: Potential carbohydrate requirement of cotton bolls. In Modeling the growth of trees, Proc. of a workshop on tree growth dynamics and modeling. ed. Murphy et al. 123-127. Duke University.

Hutmacher, R. B., D. R. Krieg, C. J. Phene, S. S. Vail, and L. H. Ziska. 1986. Leaf conductance and growth responses to vapor pressure deficit and temperature: cotton, cowpea, and tomato. Agronomy Abstract. 97.

Jordan, W. R. 1983. Cotton. In Crop-water relation. eds. I. D. Teare and M. M. Peets. New York: John Wiley & Sons.

Kennedy, C. W., M. T. Ba, A. G. Cladwell, R. L. Hutchinson, and J. E. Jones. 1987. Differences in root and shoot growth and soil moisture extraction between cotton cultivars in an acid subsoil. Plant and Soil. 101:241-246.

Klepper, B., H. M. Taylor, M. G. Huck, and E. L. Fiscus. 1973. Water relations and growth of cotton in drying soil. Agronomy Journal. 65:307-310.

Lety, J., L. H. Stolzy, G. B. Blank and O. R. Lunt. 1961. Effect of temperature on oxygen diffusion rates subsequent shoot growth and mineral content of two plant species. Soil Science. 92:314-321.

Mauney, J. R. and L. L. Phillips. 1963. Influence of daylength and night temperature on flowering in Gossypium. Botanical Gazette. 124:278-283.

Mauney, J. R. 1968. Morphology. In Advances in production and utilization of quality cotton. Principles and practices. eds. F. C. Elliot, M. Hoover, and W. K. Porter, Jr. Ames, IA: Iowa State University Press.

Mauney, J. R. 1966. Floral initiation of upland cotton (Gossypium hirsutum L.) in response to temperatures. Journal of Experimental Botany. 17:452-459.

Mauney, J. R. and J. M. Stewart. 1986. eds. Cotton Physiology. Memphis, TN: The Cotton Foundation.

Mauney, J. 1986. The effect of temperature on floral initiation and production of 4-Bract Squares by DPL-90.

In _Beltwide cotton production research conference, 1986 proceedings_. 63.

McMichael, B. L., W. R. Jordan, and R. D. Powell. 1973. Abscission processes in cotton: induction by plant water deficits. _Agronomy Journal_. 65:202-204.

McMichael, B. L. and C. D. Elmore. 1976. The effects of plant water status on cotton boll growth and water relations. In _Beltwide cotton production research conferences, 1986 proceedings_. 80.

McMichael, B. L. 1980. Unpublished results.

McMichael, B. L., J. J. Burke, J. D. Berlin, J. L. Hatfield, and J. E. Quisenberry. 1985. Root vascular bundle arrangements among cotton strains and cultivars. _Environmental Exp. Botany_. 25(1):23-30.

McMichael, B. L., and J. E. Quisenberry. 1986. Variability in lateral root development and branching intensity in exotic cotton. _Beltwide cotton production research conference proceedings_, ed. T. C. Nelson. 89. Memphis: National Cotton Council of America.

McMichael, B. L. 1988. Unpublished results.

Neilson, K. F. and R. K. Cunningham. 1964. The effect of soil temperature and form and level of N on growth and chemical composition of Italian rye grass, _Proceedings Society of Soil Scientists of America_. 28:213-218.

Nelson, L. E. 1967. Effect of root temperature variation on growth and transpiration of cotton (_Gossypium hirsutum_ L.) seedlings. _Agronomy Journal_. 59:391-395.

Pearson, R. W. 1974. Significance of rooting pattern to crop production and some problems of root research. In _The plant root and its environment_. ed. E. W. Carson. 247-270. Charlottesville: University of Virginia Press.

Pearson, R. W. and Z. F. Lund. 1968. Direct observation of cotton root growth under field condition. _Agronomy Journal_. 60:442-443.

Quisenberry, J. E., W. R. Jordan, B. A. Roark, and D. W. Fryrear. 1981. Exotic cotton as genetic sources for drought resistance. _Crop Science_. 21:889-895.

Ritchie, J. T. and E. Burnett. 1971. Dryland evaporation flux in a subhumid climate: II. Plant influences. _Agronomy Journal_. 63:56-62.

Russell, R. S. 1977. _Plant root systems_. New York: McGraw-Hill.

Sachs, T. and K. V. Thimann. 1967. The role of auxins and Cytokinins in the release of buds from dominance. _American Journal of Botany_. 54:136-144.

Smucker, J. M., S. L. Mcburney and A. K. Srivastava. 1982. Quantitative separation of roots from compacted soil profiles by the hydropneumatic elutriation system. Agronomy Journal. 74:500-503.

Stapleton, H. N., D. R. Buxton, Y. Makki, and R. E. Briggs. 1973. Some effects of field weathering of seed cotton in a desert environment. Agronomy Journal. 65:14-15.

Taylor, H. M. and H. R. Gardner. 1963. Penetration of cotton seedling taproots as influenced by bulk density, moisture content, and strength of soil. Soil Science. 96:153-156.

Taylor, H. M. and L. F. Ratliff. 1969. Root elongation rates of cotton and peanuts as a function of soil strength and soil water content. Soil Science. 108:113-119.

Taylor, H. M. and B. Klepper. 1974. Water relations of cotton. I. Root growth and water use as related to top growth and soil water content. Agronomy Journal 66:584-588.

Taylor, H. M. and B. Klepper. 1978. The role of rooting characteristics in the supply of water to plants. Advances in Agronomy. 30:99-128.

Taylor, H. M. 1983. Managing root systems for efficient water use: an overview. In Limitations to efficient water use in crop production. eds. H. M. Taylor, W. R. Jordan, and T. R. Sinclair. 87-113. Madison, WI: American Society of Agronomy.

Taylor, R. M. 1972. Germination of cotton (Gossypium hirsutum L.) pollen on an artificial medium. Crop Science. 12:243-244.

Tharp, W. H. 1965. The cotton plant, how it grows, and why its growth varies. Agriculture Handbook No. 178. USDA-ARS.

Tollervey, F. E. 1970. Physiology of the cotton plant. Cotton Growing Review. 47:245-256.

Torrey, J. G. 1976. Root hormones and plant growth. Annual Review Plant Physiology. 27:435-459.

Upchurch, D. R., B. L. Michael, and H. M. Taylor. 1988. Use of minirhizotrons to characterize root system orientation. Soil Science Society American Journal. 52:319-323.

Wanjura, D. F. and D. R. Buxton. 1972. Water uptake and radicle emergence of cottonseed as affected by soil moisture and temperature. Agronomy Journal. 64:427-431.

Wullschleger, S. D. and D. M. Oosterhuis. 1989. Photosynthesis of individual field-grown leaves during omtogeny. Photosynthesis Research (in press).

Randy Moore

11 The Effects of Gravity on the Ecology and Dynamics of Root Growth

Gravity is the most constant and ubiquitous environmental stimulus that influences evolution and growth of plants. Botanists have tried to understand gravity's influence on plant growth since before the time of Charles Darwin. We now know that gravity affects several aspects of root growth and development, including mineral nutrition and the direction of growth (see reviews by Halstead and Dutcher 1987; Moore 1988). The ubiquitous nature of gravity, however, has made its precise influence almost impossible to determine. Designing a control that eliminates only gravity from the experimental design has been problematic, since there are no places on Earth where we can escape gravity's pull. Some botanists have tried to "compensate" for gravity by growing plants on clinostats, in which seedlings are continually rotated horizontally with no fixed position relative to gravity. Clinostats do not abolish gravity, however; rather, seedlings rotated on clinostats are constantly exposed to 1 g, albeit from an ever-changing direction. Furthermore, rotation on clinostats mechanically stresses plants, which explains why seedlings grown on clinostats have different physiological properties than those grown in microgravity (Brown and Chapman 1985).

The importance of outer space as the only true control for gravity-related experiments is evident because of our inability to extinguish gravity on Earth. Since microgravity is not easily accessible, our knowledge of gravity's effects on plant growth and development is limited and often based on small samples lacking necessary controls. For example, the earliest experiments flown in outer space were reexposed to 1 g for several hours while the spacecraft re-entered Earth's gravitational field and materials were processed after landing. This re-exposure to gravity sometimes compromised the experiment's results. With continued flights into outer space, however, came improved experiments. For example, recent experiments have involved launching imbibed seeds and seedlings into outer space, allowing them to grow for several days, and then fixing the seedlings before re-entry into Earth's gravity (Volkmann et al. 1986; Moore et al. 1987). With such improvements in experimental protocol have come a somewhat better understanding of how gravity affects growth and development of plants. The influence of gravity on shoots and the results of Earth-based experiments are presented elsewhere (see reviews by Evans et al. 1986; Halstead and Dutcher 1987).

INFLUENCE OF GRAVITY ON GROWTH
AND INITIATION OF ROOTS

Gravity strongly affects growth and initiation of roots. For example, roots of corn (Zea mays) seedlings grown in microgravity for 4.8 days elongated 3.11 ± 0.7 cm, which was less than half that of Earth-grown controls (i.e., 7.57 ± 0.9 cm; Moore et al. unpublished results). Thus, gravity stimulates root elongation. Interestingly, shoots of flight-grown seedlings elongated 6.12 ± 1.4 cm during the same period, which was more than twice that of Earth-grown controls (2.56 ± 0.6 cm; Moore et al. unpublished results). These results suggest that gravity inhibits shoot elongation and causes plants to allocate their resources for growth preferentially to roots rather than shoots.

We have known for more than a century that primary roots of most plants are positively gravitropic (see reviews by Moore and Evans 1985; Evans et al. 1986). External stimuli such as gravity and light strongly affect growth and orientation of roots. When these stimuli are extinguished (e.g., in dark growth chambers in microgravity), the shape and orientation of the embryo determine the direction of root growth (Volkmann et al. 1986; see review by Halstead and Dutcher 1987). Interestingly, neither light nor nutrients significantly affect root growth in microgravity. This suggests that root growth is affected more by gravity than by light or nutrients. Furthermore, initiation of root growth during seed germination is not affected by microgravity; numerous experiments have demonstrated that germination rates are similar in Earth-grown controls and microgravity (see review by Halstead and Dutcher 1987).

Gravity also affects initiation of secondary roots. For example, two-day-old seedlings of Zea mays launched into microgravity produced 36.8 ± 4.9 secondary roots, while Earth-grown controls produced only 10.0 ± 1.3 secondary roots (Moore et al. unpublished results). These results indicate that gravity inhibits initiation of secondary roots. If growth regulators such as auxin and cytokinin control initiation of lateral roots (Gregory et al. 1987), then microgravity may stimulate initiation of secondary roots by altering the levels of these substances.

INFLUENCE OF GRAVITY ON STRUCTURE AND FUNCTION OF CELLS IN ROOTS

More than a century ago, Charles Darwin wrote that "there is no structure in plants more wonderful, as far as its functions are concerned, than the tip of the radicle" (Darwin 1880). The thimble-shaped root cap covering the tip of a root has several important functions, including producing mucilage and sloughing cells to help the root force its way through soil, producing effectors that affect root elongation, detecting gravity, and controlling gravitropism (Gregory et al. 1987; Moore and Evans 1986; Evans et al. 1986). Not surprisingly, gravity has many effects on root caps.

Regeneration of Root Cap

Decapped roots grown on Earth completely regenerate their caps within two to three days after decapping. However, decapped roots growing in outer space do not regenerate their caps within 4.8 days, suggesting that gravity is necessary for regeneration of root caps (Moore et al. 1987b).

Cellular Differentiation

Cells of root caps of seedlings grown in microgravity usually allocate more volume to lipid bodies and hyaloplasm and less volume to mitochondria, dictyosomes, and plastids than those of Earth-grown seedlings. Volumes of nuclei in cells of root caps are unaffected by microgravity (Moore et al. 1987). These effects of microgravity on cellular structure and differentiation are specific for particular organelles and cell types, and extend to other parts of the root. For example, roots growing in microgravity have a smaller elongating zone than Earth-grown controls (Merkeys et al. 1976). These altered patterns of cellular differentiation presumably underlie differences in characteristics of flight-grown plants, such as variations in growth and in formation of lateral roots.

Cellular Polarity

Many types of plant cells distribute their organelles polarly. Columella cells in root caps, for example, are the putative graviperceptive cells and contain large amyloplasts that sediment to the lower side of the cells (see review by Moore and Evans 1986). This rearrangement of amyloplasts (and the resulting interactions with other cellular structures such as endoplasmic reticulum) is believed by many to be how roots respond to gravity (Evans et al. 1986). However, amyloplasts of columella cells of flight-grown seedlings are distributed randomly (Moore et al. 1987). Similarly, endoplasmic reticulum of flight-grown seedlings is distributed significantly differently from that of Earth-grown seedlings (Moore et al. 1987). These observations suggest that the polar distribution of the organelles in cells of Earth-grown seedlings is controlled primarily by gravity. The loss of cellular polarity in microgravity may underlie the altered growth of roots in microgravity.

The cellular and subcellular effects of microgravity have important implications for root growth and development. Dictyosomes in peripheral cells of root tips, for example, produce and secrete mucigel, a hydrated polysaccharide containing sugars, organic acids, vitamins, enzymes, and amino acids. The amounts of mucigel produced by roots often reach astounding proportions. One hectare of corn produces more than 1,000 m^3 of mucigel, a volume roughly equivalent to that of a typical, two-story, four-bedroom house. This mucigel has several important functions, including lubrication, protection, and absorption of water and nutrients. Carboxylic acids in mucigel influence uptake of ions, and organic acids chelate several ions.

Furthermore, fatty acids and lectins may influence establishment of important symbioses. Nitrogen-fixers such as
Azospirillum grow in mucigel of plants such as corn, and in
other plants, mycorrhizal infection often parallels the
rate of mucigel exudation (Gregory et al. 1987; Russell
1977; Curl and Truelove 1986).

Gravity drastically reduces the number and relative
volume of dictyosomes in cells of root tips that produce
mucigel (Moore et al. 1987). For example, cells of plants
grown in microgravity allocate 93 percent less volume to
dictyosomes than do cells of Earth-grown controls (Moore et
al. 1987). Since dictyosomes produce and secrete mucigel,
it is not surprising that flight-grown plants secrete significantly less mucigel than Earth-grown controls. Since
mucigel is important for a plant's mineral nutrition, these
results suggest that gravity significantly alters secretion
and, as a result, nutrition. Plants grown in microgravity
contain more phosphorous and potassium, and less calcium,
magnesium, manganese, zinc, and iron than Earth-grown controls (Halstead and Dutcher 1987). These differences in
mineral nutrition are probably due to altered absorption by
roots in microgravity.

Cell walls are a plant's primary means of withstanding
gravity. Since cell walls are deposited by dictyosomes,
the reduced allocation of dictyosomes in cells of roots
also affects the deposition and structure of cell walls.
Walls of flight-grown seedlings are thinner and contain
less lignin and cellulose than those of Earth-grown controls (Cowles et al. 1984, 1986). These results suggest
that gravity alters the amount and composition of cell
walls produced by plant cells.

Cellular and subcellular studies of the effects of
gravity on roots also suggest that gravity affects the
energy metabolism of roots. For example, most cells in
roots store energy as starch in amyloplasts. Amyloplasts
of flight-grown seedlings are smaller and contain smaller
starch-grains than Earth-grown controls (Moore et al.
1986). These results suggest that gravity affects how
plants metabolize starch, and thus have important implications for NASA's long-range plans for growing crops in
outer space. These results predict that starch-storing
organs such as potatoes grown in microgravity would contain
less starch and be of less caloric value than Earth-grown
controls.

Cells of roots grown in microgravity contain significantly more lipid bodies than Earth-grown controls (Moore
et al. 1986). This observation, combined with the fact

that roots of flight-grown plants contain less starch than Earth-grown controls, suggests that microgravity alters energy-metabolism in roots by shifting energy storage from starch to lipid.

FUTURE PROSPECTS

Botanists must continue to exploit microgravity to understand how gravity affects growth and development of roots. Although short flights limit experiments performed on the space shuttle, deploying a space station should enable us to design long-term experiments investigating phenomena such as development of root systems, root-shoot partitioning, and the influence of microgravity on reproduction and food storage.

REFERENCES

Brown, A. H., and D. K. Chapman. 1985. How far can clino-
    stats be trusted? Physiologist 28:297.
Cowles, J., H. W. Scheld, R. LeMay, and C. Peterson.
    1984. Experiments on plants grown in space: Growth
    and lignification in seedlings exposed to eight days
    in microgravity. Annals of Botany 54(Suppl.
    30):33-48.
Cowles, J. R., R. LeMay, R. Omran, and G. Jahns. 1986.
    Cell wall related synthesis in plant seedlings grown
    in the microgravity environment of the space shuttle.
    Plant Physiology 80(Suppl. 4):9.
Curl, E. A., and B. Truelove. 1986. The Rhizosphere.
    Advanced Series in Agricultural Sciences no. 15.
    Berlin: Springer-Verlag.
Darwin, C. 1880. The Power of Movement of Plants.
    London: J. Murray.
Evans, M. L., R. Moore, and K. Hasenstein. 1986. How
    roots respond to gravity. Scientific American
    255:112-119.
Gregory, P. J., J. V. Lake, and D. A. Rose, eds. 1987.
    Root development and function. Society for
    Experimental Biology, Seminar Series no. 30.
    Cambridge, England: Cambridge University Press.
Halstead, T. W., and F. R. Dutcher. 1987. Plants in
    space. Annual Review of Plant Physiology 38:317-345.
Merkys, A. J., R. S. Laurinavichius, A. L. Mashinskiy, A.
    V. Yaaaroshius, E. K. Savichene, et al. 1976. Effect
    of weightlessness and its simulation on the growth and
    morphology of cells and tissues of pea and lettuce
    seedlings. In Organizmy i sila tyazhesssti.
    materialy i vsesoyuznoy konferentsii "gravitatsiya i
    organism", ed. A. J. Merkys, 238-246. Vilnius:
    Institute of Botany, Lithuanian Soviet Socialist
    Republic Academy of Sciences and Institute of General
    Genetics, Union of Soviet Socialist Republic Academy
    of Sciences (In Russian).
Moore, R., and M. L. Evans. 1986. How roots perceive and
    respond to gravity. American Journal of Botany
    73:574-587.
Moore, R., W. M. Fondren, C. E. McClelen, and C-L. Wang.
    1987. Influence of microgravity on cellular
    differentiation in root caps of Zea mays. American
    Journal of Botany 74:1006-1012.

Moore, R., C. E. McClelen, W. M. Gonddren, and C-L. Wang. 1987. Influence of microgravity on root-cap regeneration and the structure of columella cells in Zea mays. American Journal of Botany 74:218-223.

Moore, R., W. Mark Fondren, E. C. Koon, and C. L. Wang. 1986. The influence of microgravity on the formation of amyloplasts in columella cells of Zea mays L. Plant Physiology 82:867-868.

Russell, R. S. 1977. Plant root systems: The function and interaction with the soil. London: McGraw-Hill.

Volkmann, D., H. M. Behrens, and A. Sievers. 1986. Development and gravity sensing of cress roots under microgravity. Natur-wissenschaften 73:438-441.

M. C. Whalen
L. J. Feldman

12 Effects of Mechanical Impedance on Root Development

External pressure physically altering development has been the historical explanation for the response of roots to mechanical impedance. The range and variability of responses to impedance are too complex, however, to be explained by pressure alone. Growth substances are now believed to mediate these responses. This chapter will review the current knowledge about the role of growth substances, focusing on the the role of ethylene.

The growth of roots in soil depends on a supply of energy-rich compounds from the shoot or storage organ; the availability of water, oxygen, and mineral nutrients; the temperature; and the physical properties of the soil (Eavis and Payne 1969; Hillel 1980, p. 413).

Sustained root growth results from the production of new cells by the apical meristem and from cell enlargement. Turgor pressure drives cell enlargement and hence elongation of the entire root. For growth in soil, the turgor pressure must be great enough to overcome the resistance of the soil (Taylor et al. 1966; Taylor 1974). In the late nineteenth century, researchers found that roots generated pressures ranging from 5 to 13 bars (Pfeffer 1893). These considerable pressures were believed to be important in determining whether roots were able to grow successfully in a variety of soil environments. Pressures generated by roots are particularly important for penetration in compacted soils, where there are few pores to allow unimpeded root penetration. At some level of soil compaction or soil strength, roots become mechanically stressed and penetration decreases. When other factors for growth are satisfactory, root growth in compacted soil decreases as the soil strength increases (Voorhees et al. 1975).

Besides a reduction in the rate of growth (Russell and Goss 1974), impeded roots often show other developmental changes. Impeded roots are wider than nonimpeded roots, and first- and second-order laterals proliferate uncharacteristically close to the tip of the root (Goss 1977; Goss and Russell 1980). Although elongation is inhibited, the volume of root matter per unit of soil is greater because of the production of first-order laterals (Voorhees et al. 1975).

Because of its implications for agricultural productivity and technical practices, the mechanism by which mechanical impedance affects root growth in the field is of considerable interest. Studies in the field, however, are hindered by the inability to control experiments done in soil (Russell and Goss 1974). Many experimental apparatuses have been constructed to attempt to overcome this difficulty. With these apparatuses, the pressure or mechanical stress applied to a root may be precisely regulated, while other factors are held constant. Using such devices, investigators have shown that root development is sensitive to applications of low pressures. Russell and Goss (1974) showed that at an applied pressure of 0.15 bars, the elongation of barley roots was only 75 percent of the elongation of controls. At a higher pressure of 0.5 bars, elongation was 20 percent that of controls. At both pressures, reduced elongation was accompanied by thickening of the root. Use of such experimental devices has also shown that other environmental factors, such as oxygen availability, greatly influence the sensitivity of roots to mechanical impedance (Gill and Miller 1956; Barley 1962).

The physiological basis for the response of roots to mechanical impedance is not known. Barley (1962), and later Gracean and Oh (1972), explained the effect of mechanical impedance in terms of osmotic adjustment. They suggested that since cell length in mechanically impeded roots was reduced, mechanical stress directly reduced the final volume of cells. This view is incorrect, because the volume of cells in impeded roots is often greater than that of cells in nonimpeded roots.

Another line of inquiry is based on observations that the cells composing or adjacent to the root apical meristem may be particularly important in detecting and transducing applied pressure. Extension was considerably reduced when intact maize roots were forced to collide with small glass beads. In roots from which the cap had been excised before contact with the beads, there was no significant change in elongation rates (Goss and Russell 1980). The effect of decapping alone on elongation rates, however, was not taken into account in these experiments. It is therefore unclear whether the root cap truly mediates responses to mechanical resistance. Further work is necessary to determine whether the root cap transduces physical stress into a physiological response. Other tissues within the root meristem are also likely to participate in the response to mechanical impedance. Depending on the species, one to three days is necessary before one can observe the full inhibitory response to mechanical impedance (Barley 1962; Taylor and Ratliff 1969). Following release of pressure, extension rates do not return to control values for two to three days (Goss and Russell 1980). The kinetics of elongation following application of or release from applied pressure suggest that the response to mechanical impedance is dependent on cellular activities within the apical meristem.

Initially the observed patterns of root response to mechanical impedance were explained as external pressure directly reducing the rate and degree of cell enlargement. It is now clear that the range and variability of responses of roots to mechanical impedance are too complex to be explained by pressure alone (Feldman 1984). A few reports advanced the idea that growth substances may mediate impedance responses in roots (Kays et al. 1974; Lachno et al. 1982). Goss and Drew (1972) reported that when a mechanically impeded root was forced to bend, lateral roots usually developed on the convex side of the bending root. Since lateral root initiation is postulated to be under hormonal control (Torrey 1976), the work of Goss and Drew suggests that mechanical impedance may affect the synthesis of growth regulators.

To address the interaction between mechanical impedance and plant growth substances, researchers mechanically impeded plant tissues and monitored concentrations of growth regulators (Goeschl et al. 1966; Kays et al. 1974; Lachno et al. 1982). Developmental changes observed in response to mechanical impedance were then correlated with the observed changes in concentrations of growth substances. In a study by Lachno, Harrison-Murray, and Audus (1982), maize roots were grown in compacted soil and levels of abscisic acid (ABA) and indole-3-acetic acid (IAA) were measured in root tips. These workers found that the concentration of ABA was not affected by compaction, but that levels of IAA were 3.5 times greater in impeded roots than in controls. The increased levels of auxin produced in response to mechanical impedance may directly or indirectly inhibit root growth (Lachno et al. 1982). Since auxin stimulates ethylene production in vegetative tissues (Chadwick and Burg 1970; Yoshii and Imaseki, 1981; Yu et al. 1979), auxin may exert its primary influence on impeded roots through ethylene.

In another study, Kays, Nicklow, and Simons (1974) grew roots of broad bean in perforated plastic tubes with an impenetrable plug at one end. Measuring rates of ethylene evolution, they found that the rate in impeded roots was six times greater than that in controls. When the obstruction to downward growth was removed, rates of ethylene evolution decreased to control levels. The inhibition of root elongation in response to mechanical impedance was similar to inhibition by exogenous ethylene, and the researchers therefore suggested a causal relationship.

Goeschl, Rappaport, and Pratt (1966) also addressed the question of whether ethylene was primarily responsible for the response of plant tissues to mechanical impedance. They grew pea epicotyls in columns of glass beads and found that ethylene evolution increased greatly in response to this physical resistance. Developmental changes induced by growth in beds of beads were correlated with those induced by exogenous ethylene. It is apparent from these studies that ethylene may be the controlling factor in the growth response to mechanical impedance.

Knowing from our work that ethylene-induced changes in maize root development are similar to the changes induced by mechanical impedance, we investigated whether ethylene was involved in the response to mechanical impedance (Whalen 1988; Whalen and Feldman 1988). To clarify the role of ethylene in the response to mechanical impedance, we first refined the experimental methods of imposing mechanical impedance. Then we studied the kinetics and biochemistry of ethylene production in response to short-term exposures.

To increase substantially our understanding of the early physiological responses to mechanical impedance, it was necessary to experimentally isolate the axial from the lateral components of mechanical impedance. This was important, considering the work of Biro and Jaffe (1984) showing that mechanical perturbation affects ethylene metabolism. Their work suggests that growing plant tissue in beds of beads or in plastic tubes may affect ethylene metabolism independently of an effect from obstructing downward growth. By not restraining roots in beds of beads or in plastic tubes and by monitoring responses immediately after roots collided with an obstruction to downward growth, we experimentally isolated the axial from the lateral components of mechanical impedance. We designed a plexiglas chamber with which we could impede twenty-eight seedling roots of maize and measure ethylene gas (Whalen 1988). The experimental design allowed a single event of axial impedance to take place, rather than repeated axial and lateral events. It is difficult to control the time required for ethylene gas to diffuse from plant tissues and accumulate to detectable levels. Study of the precise timing of the onset of changes in ethylene metabolism by measuring external ethylene levels is therefore often inaccurate. To reduce inaccuracy we monitored the concentration of the biosynthetic precursor of ethylene, 1-aminocyclopropane-1-carboxylic acid (ACC). This concentration reflects the potential of tissues to produce ethylene (Yang and Hoffman 1984).

Within 300 minutes after roots collided with the obstruction to downward growth, ethylene evolution decreased. Interestingly, although impedance inhibited ethylene evolution, it stimulated ACC production. ACC concentrations in impeded roots were well within the levels that are normally converted to ethylene (Whalen 1988). It appears that the biochemical conversion of ACC to ethylene was inhibited in some fashion by impedance. Our results that ethylene evolution was inhibited in response to impedance are at variance with the results of other workers (Goeschl et al. 1966; Kays et al. 1974). Three factors may be responsible for these discrepancies: (1) the combined effect of both lateral and axial forces exerted throughout an experiment may be necessary for the reported stimulation of ethylene; (2) there may be differences in the responses to impedance of monocot and dicot roots; (3) and the length of time plant tissues are impeded may be important. Indeed, concerning this last point, early responses to mechanical impedance do appear to be different from later responses.

Since the root cap is considered to be important in the growth response of roots to mechanical impedance (Goss and Russell 1980), rates of ethylene evolution from decapped impeded roots were compared with rates from decapped control roots. Impedance inhibited ethylene evolution by decapped roots to the same extent as intact roots. We conclude that the root cap is not required for the observed early response of ethylene metabolism to impedance (Whalen 1988).

Although much has been learned about the physiology of the developmental response of roots to mechanical impedance, much remains unanswered. Because of experimental constraints, studies of the responses of roots to physical stresses are difficult to perform. Thus far it has not proved possible to uncover definitive causal relationships. Genetic mutants and inhibitors of plant growth substances have not yet been used in studies of mechanical impedance. In other areas of study, these two experimental tools have had a high degree of success in allowing definition of causal relationships. Further development of these tools, plus efforts to design better devices for imposing physical stress, may lead to a more complete understanding of the physiological basis of the response of root growth to mechanical impedance.

REFERENCES

Barley, K. P. 1962. The effects of mechanical stress on the growth of roots. Journal of Experimental Botany 13:95-110.

Biro, R. L., and M. J. Jaffe. 1984. Thigmomorphogenesis: ethylene evolution and its role in the changes observed in mechanically perturbed bean plants. Physiologia Plantarum 62:289-296.

Chadwick, A. V., and S. P. Burg. 1970. Regulation of root growth by auxin-ethylene interaction. Plant Physiology 45:192-200.

Eavis, B. W., and D. Payne. 1969. Soil physical conditions and root growth. In Proceedings of the fifteenth Easter school in agricultural science, ed. W. J. Butterworth, 315-338. London: Whittington.

Feldman, L. J. 1984. Regulation of root development. Annual Review of Plant Physiology 35:223-242.

Gill, W. R., and R. D. Miller. 1956. A method for study of the influence of mechanical impedance and aeration on the growth of seedling roots. Soil Science Society of American Proceedings 20:154-157.

Goeschl, J., L. Rappaport, and H. K. Pratt. 1966. Ethylene as a factor regulating the growth of pea epicotyls subjected to physical stress. Plant Physiology 55:670-677.

Goss, M. J. 1977. Effects of mechanical impedance on root growth in barley (Hordeum vulgare). I. Effects on elongation and branching of seminal root axis. Journal of Experimental Botany 28:96-111.

Goss, M. J., and M. C. Drew. 1972. Effect of mechanical impedance on growth of seedlings. Agricultural Research Council, Letcombe Laboratory Report 1971:35-42.

Goss, M. J., and R. S. Russell. 1980. Effects of mechanical impedance on root growth in barley (Hordeum vulgare L.). III. Observation on the mechanism of response. Journal of Experimental Botany 31:577-588.

Gracean, E. L., and J. S. Oh. 1972. Physics of root growth. Nature New Biology 235:24-25.

Hillel, D. 1980. Fundamentals of soil physics. New York: Academic Press.

Kays, S. J., C. W. Nicklow, and D. H. Simons. 1974. Ethylene in relation to the response of roots to physical impedance. Plant and Soil 40:565-571.

Lachno, D. R., R. S. Harrison-Murray, and L. J. Andus.

1982. The effects of mechanical impedance to growth on the levels of ABA and IAA in roots tips of Zea mays. _Journal of Experimental Botany_ 33:943-951.

Pfeffer, W. 1893. Druck-und Arbeits-leistung durch wachsende Pflazen. Abhandlungen der Sachsischen Akademie der Wissenschaften, _Mathematichsch-Physische_ 33:235-474.

Russell, R. S., and M. J. Goss. 1974. Physical aspects of soil fertility--the response of roots to mechanical impedance. _Netherlands Journal of Agricult_ 22:305-318.

Taylor, H. M. 1974. Root behavior as affected by soil structure and strength. In _The plant root and its environment_, ed. E. W. Carson, 271-291. Charlottesville: University of Virginia Press.

Taylor, H. M. and L. F. Ratliff. 1969. Root elongation rates of cotton and peanut as a function of soil strength and soil water content. _Soil Science_ 108:113-119.

Taylor, H. M., G. M. Robertson, and J. J. Parker, Jr. 1966. Soil strength-root penetration relations for medium tocoarse textured soil materials. _Soil Science_ 102:18-22.

Torrey, J. G. 1976. Root hormones and plant growth. _Annual Review of Plant Physiology_ 27:435-459.

Voorhees, W. B., D. A. Farrell, and W. E. Larson. 1975. Soil strength and aeration effects on root elongation. _Soil Science Society of America Proceeding_ 39:948-953.

Whalen, M. C. 1988. The effect of mechanical impedance on ethylene production by maize roots. _Canadian Journal of Botany_ 66:2139-2142.

Whalen, M. C., and L. J. Feldman. 1988. The effect of exogenous ethylene on root growth of Zea mays seedlings. _Canadian Journal of Botany_ 66:719-723.

Yang, S. F., and N. E. Hoffman. 1984. Ethylene biosynthesis and its regulation in higher plants. _Annual Review of Plant Physiology_ 35:155-189.

Yoshii, H. and H. Imaseki. 1981. Biosynthesis of auxin-induced ethylene. Effects of indole-3-acetic acid, benzyladenine, and abscisic acid on endogenous levels of ACC and ACC synthase. _Plant Cell Physiology_ 22:369-379.

Yu, Y. B., D. O. Adams, and S. F. Yang. 1979. Regulation of auxin-induced ethylene production in mung bean hypocotyls: role of ACC. _Plant Physiology_ 63:589-590.

Morris G. Huck
Gerrit Hoogenboom

13 Soil and Plant Root Water Flux in the Rhizosphere

Movement of water into and through the soil profile is essential for life on Earth as we know it. Water held in soil storage is a fundamental requirement for the growth of plants on dry land, and thus it is an essential component of the food chains for most terrestrial animal life. The biochemical processes of all organisms have a narrowly defined range of water potential for maintaining functional and structural integrity of critical enzymes. Because the internal water storage capacity of most plants is minimal compared to the amount lost from their leaves between rainfall events, soil water storage is the key to survival.

Water entry and movement in the soil profile is a complex process that depends upon both soil properties and the availability of water. Water from rainfall or irrigation enters the soil profile at the surface and, under the influence of gravity, gradually flows downward through various soil horizons or layers. When water is applied faster than it can be absorbed by the soil, it accumulates at the surface, where it may be lost by evaporation or surface runoff. As surface-applied water gradually percolates through the soil, it replaces air held in the soil pore spaces. Excess water, beyond that which can be held in the soil profile, is gradually lost by deep percolation and drainage into the ground water or water table.

Soil hydraulic properties, which influence the flow of water through soil, are affected by the size and packing density of soil particles, as well as by irregularities in soil structure such as cracks or channels from old roots and soil fauna. Water is removed from soil storage by plant root systems, whose growth is very sensitive to micro-environmental conditions within the soil. Roots grow into soil regions where water reserves are available, often at a rate exceeding that at which liquid water films move through unsaturated soil. Moreover, the growth of roots in a soil profile exerts a strong influence on soil porosity and the rates of infiltration and percolation.

Water adhering to the surface of soil particles forms a lubricating film that greatly reduces the effective mechanical strength. Thus, even small amounts of water in the soil can act as a lubricant, permitting soil particles to slip past each other and reducing the utility of a soil as a mechanical support medium. For ideal root growth and function, then, a soil should be loose and friable so that roots can easily penetrate, and so that water can readily percolate through the profile (Bristow, Campbell, and Calissendorff 1984; Goss et al. 1984; Hamblin 1985).

A loose, friable soil that facilitates water percolation and root growth is ideal for plant growth. With the advent of modern society, non-agricultural land uses have evolved different sets of desirable physical characteristics for soils. Foundations for highways and buildings must be mechanically rigid to support heavy loads under all kinds of conditions. Thus, the primary objective of highway and building construction engineers is best achieved when soil particles are densely packed into a mechanically rigid matrix (Craig 1983). Water must be excluded from foundations and roadbeds, for example, if these structures are to perform properly. Paving upon the surface of highways, airports, and parking lots is designed to prevent infiltration of rain water. An understanding of soil water relations, therefore, is as important for construction engineers as for members of the agricultural community.

SOIL WATER STORAGE

Before discussing water movement, we must consider the physical properties of the soil medium: the forms of soil water storage, the nature of the soil pores through which water moves, and the characteristics of plant root systems and how they influence soil structure. To understand the soil-plant-atmosphere system fully, we must "dissect" it, but we must also remember that our purpose in analyzing each part of the system is to gain an understanding of its role in the functioning of the complete system (Millington and Peters 1970). Because professionals in many different disciplines study water movement in soils (e.g., agricultural engineers, soil physicists, biophysicists, crop physi-

ologists, biochemists, and agrometeorologists), it is
essential that the terms used be understood by all.

Soil Structure and Pore Spaces

Most natural soils are composed of finely divided bits
of minerals, which are irregularly shaped and separated by
small pore spaces. This solid particulate matter is generally mixed together in a somewhat random fashion by the
natural soil-forming processes of weathering, erosion, and
deposition, or by the action of biotic agents such as
growing roots or soil microfauna. Since the individual
soil particles do not fit together in a regular lattice (as
is characteristic of the crystalline structure of their component minerals), they are separated by small pore spaces.
It is in these pore spaces, between the irregularly shaped
mineral soil particles, that water and air are held--and
through which soil water movement occurs, in both liquid
and vapor phases.

The closer that individual soil particles are packed
together, the higher the bulk density, or mass per unit
soil volume. Freshly disturbed soil material (depositions
of water-borne sediment, loess, or even recently tilled
soil) is relatively loose because of the random orientation
of the soil particles. With time and the action of pressure from overburden or by traffic from animals and machines, the soil particles are packed together more closely
(Khatibu, Lal, and Jans 1984). As bulk density increases
(with tighter packing of individual soil particles), the
pore space remaining between particles (containing air,
water films, growing roots, and associated microbiota) will
be reduced (Gupta and Larson 1979a), and root growth will
be significantly decreased (Thompson, Jansen, and Hooks
1987). The characteristic soil water retention curve (the
relationship between soil water content and the potential
energy with which water is held) and other hydraulic properties of the soil are also modified as soil porosity is
reduced by compaction (Gupta and Larson 1979b; Klute 1982).

Over the short term (days or weeks), in the absence of
mechanical forces to induce soil compaction, soil porosity
and thus bulk density tends to remain relatively constant.
Soil pore spaces, however, are filled with a mixture of
water and air. Thus as soil water content increases following rainfall or irrigation events, the volume of air in the
pore spaces is reduced; or, conversely, as water is lost
from soil storage, the volume fraction of air in the soil

pore spaces increases. Vapor-phase transport of water and other gaseous diffusion processes are facilitated as the total volume fraction occupied by air-filled pore spaces increases.

Soil Texture and Particle-Size Distribution

As the size of individual soil particles is reduced (a reduction that occurs naturally as a result of weathering and other soil-forming processes), the tendency for water to adhere gradually increases because the ratio between surface area and particle mass is an exponential function of particle size. Resistance to deformation at a given water content varies widely with particle-size distribution and packing density (Wilun and Starzewski 1972).

Mineral soil particles are grouped by size classes, although the categories are usually rather arbitrary (Hillel 1980, pp. 60-61). Several different classification schemes are in common usage. The system most often used in agriculture is that defined by the Soil Conservation Service of the U.S. Department of Agriculture (Soil Survey Staff 1975), in which particles of the smallest diameter (less than 0.002 mm) are classed as clay; intermediate-sized particles (0.002-0.05 mm diameter), as silt; coarser particles (0.05-2.0 mm), as sand; and very coarse particles (greater than 2.0 mm), as gravel.

In addition to mineral soil particles, the solid fraction of a natural soil often contains significant amounts of organic matter. These soil organic components are usually very hydrophilic, often holding several times their weight in adsorbed water. Therefore there is a very strong relationship between particle-size distribution, organic matter in the soil, and water holding capacity (De Jong, Campbell, and Nicholaichuk 1983).

SOIL WATER MOVEMENT

Van Genuchten and Wierenga (1976) proposed a system for distinguishing microscopic regions of the soil with respect to storage and flow of water (Figure 13.1). According to their analysis, the soil can be divided into an air space, a liquid space, and a solid space. Within the solid

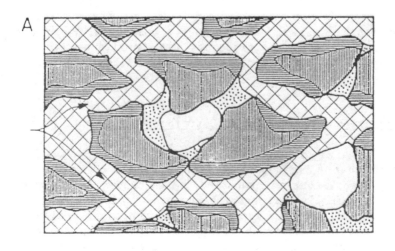

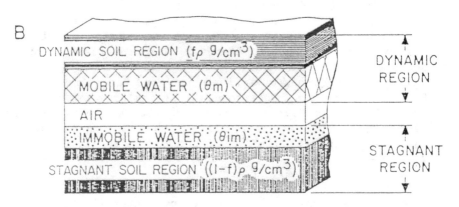

Figure 13.1. Soil particles are shown surrounded by films of capillary water and connecting irregular pore spaces. (A) actual geometry; (B) simplified model. The shading patterns in (A) and (B) represent the same regions (after van Genuchten and Wierenga 1976).

space, there is a dynamic soil region that is actually involved in adsorption of water, solute, and chemicals, and a stagnant or "dead" soil region with limited sorption. The liquid region is differentiated into an immobile region, where water forms a very tight bond with the soil particles and can only be removed by strong forces, and a mobile or dynamic region, where most of the water movement occurs.

Flow Pathways

On the macroscopic scale, water infiltrates the soil profile by percolation into the surface following rainfall or an irrigation event. Occasionally it may also enter the profile as a result of underground seepage (Swartzendruber and Hillel 1973). Not all water reaching the soil surface enters the profile. When rainfall or irrigation water arrives at the soil surface faster than it can infiltrate the profile, the excess water will either accumulate on the surface or run off via surface drainage channels. The rate of infiltration into the profile is a function of soil permeability and a characteristic of the size and contiguity of pores and cracks in the soil structure.

Water is sometimes artificially applied: farmers use various types of irrigation systems such as flooding, sprinkling the entire soil surface, or installing miniature emitters at the base of each plant (trickle or drip irrigation) to overcome drought during certain periods of the growing season in order to increase yield and production.

The more nearly level the soil surface, the more slowly excess water will run off, and thus the more time it has to soak into the soil. On more sloping soils, or whenever streams or small ditches are accessible, the accumulating surface water may run off into surface drainage channels so fast that it carries some of the soil along with it (erosion) and deposits the soil material downstream.

If free water remains ponded on the soil surface, or if the surface layers are continually wet, evaporation may occur, just as it does from a lake or any free water surface. Because evaporation requires large amounts of energy, the rate at which water evaporates from a soil or free water surface depends upon its temperature and the rate at which energy lost to evaporation can be replaced (this is called an energy balance: evaporation requires 560 calories per mole of liquid water converted to the vapor phase). Thus, to maintain a constant temperature at the soil surface, solar insolation must replace the energy lost to evaporative cooling (Jackson et al. 1973, 1974). This will occur much faster on sunny days when the soil is bare (not shaded by growing plants or other ground cover such as crop residues).

Percolation and Drainage

Once in the soil, water moves through a labyrinthine network of contiguous pore spaces. These pore spaces are the principal sites of biological activity in the soil, for here water films are in intimate contact with finely divided soil particles. Plant roots and their associated micro-organisms obtain needed water, air, and mineral resources from water films, only a few molecules in thickness, that are held in intimate contact with finely divided soil materials. A general Forrester flow diagram for water movement is given in Figure 13.2.

Water that moves into the soil continues penetrating because the soil is porous. Finally, it passes beyond the system in which we are interested, a process defined as "drainage." Often, the lower boundary of the soil profile is taken as the water table, or the point at which the soil is completely saturated with free water. If water continued to percolate into the profile and there were no loss of liquid water to underground aquifers, drainage tiles, streams and rivers, etc., the soil profile would eventually fill completely to form a swamp (as is the case in many wetland areas). Thus, drainage represents the process by which liquid water is removed from deeper profile layers or escapes from the surface soil into an underground aquifer (e.g., porous sandstone, gravel layers).

After each rainfall or irrigation event, water continues draining from the profile until the matric potential (adsorptive forces binding it to the soil particles) is just balanced by the gravitational potential (Hillel 1980). Sometimes the percolating water encounters an impermeable or slowly permeable layer in the soil profile (such as a clay lens adjacent to a sandy layer above). In this case, water above the relatively impermeable layer is said to be a "perched" water table, since it is not contiguous with the permanent water table or ground water aquifers.

At equilibrium, a significant fraction of the soil pore spaces will be filled with air (saturated with water vapor), and the gravitational potential acting to move water downward through the profile is just balanced by the adsorptive forces binding the films of water to the soil particles (matric potential). Roots and other soil-dwelling organisms have free access to both air and water films at this point (sometimes referred to in older literature as "field capacity"). The air/water mixture in the soil pore spaces at this point is nearly ideal for root growth and development.

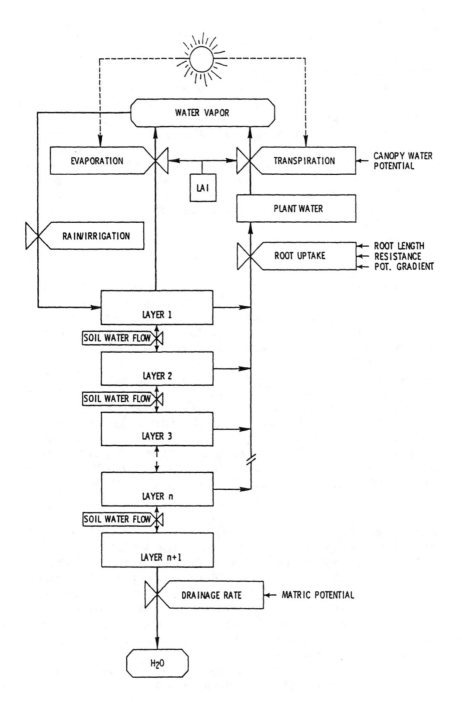

Figure 13.2. Forrester flow diagram for water balance algorithm (after Hoogenboom and Huck 1986).

Saturated Flow. Although the usual situation in most soils near the Earth's surface is for pore spaces to contain both air and water (unsaturated flow), the simpler case--when all pores are filled with a single, homogeneous fluid (saturated flow, as defined by Klute 1965)--will be considered first. Saturated flow represents a two-phase system, consisting of a stationary, solid medium (porous soil) filled with a mobile fluid (liquid water), and is nearly analogous to the situation encountered in petroleum engineering when oil or natural gas fills contiguous interstitial pores of a medium such as sandstone or other porous rock formations (Corey, Slatyer, and Kemper 1967). In simplest form, the equation for saturated flow in one dimension can be represented by Darcy's law:

$$v = Q/A = K(\Psi_2 - \Psi_1)/l \qquad [1]$$

where v equals the velocity of fluid movement in the direction of interest, Q equals the quantity of fluid moving across a given cross-sectional area, A, which is taken normal to the direction of flow; and $(\Psi_2 - \Psi_1)$ is the hydraulic potential difference between two positions in the soil, separated by a distance l.

K is a proportionality constant, which is often characterized by experimental measurement under laboratory conditions (see Marshall and Holmes 1979; Hillel 1980) and depends (among other factors) upon the particular combination of particle-size fractions constituting the soil and their packing density. Theoretical discussions of fluid flow in porous media are well documented (Brooks and Corey 1964; Baer 1969; Philip 1969; Molz and Remson 1970).

Unsaturated Flow. While the physical description of saturated flow is simpler, the more usual situation in biologically active soils is a three-phase system in which the pore spaces contain both air (saturated with water vapor) and water in the liquid phase. Liquid water films move much more slowly under unsaturated conditions because many of the pores are not contiguous. The cross-sectional area of fluid moving through a given volume of solid increases at higher water contents.

Darcy's law can also be applied to unsaturated flow, usually in differential form, expressed as

$$q = - K \frac{\delta H}{\delta x} \qquad [2]$$

where q is flux density (or flux), K is hydraulic conductivity, H is hydraulic potential, and x is flow distance. Again, experimental values for K are determined empirically and represented as a function of soil water content for a particular soil horizon.

Although the usual direction of water movement in unsaturated soil is usually downward (under the influence of gravity), upward flow can also occur (Van Bavel, Brust, and Stirk 1968) when surface evaporation or root extraction has removed water from soil layers near the surface to such an extent that total water potential ($\Psi_{gravitational} + \psi_{matric}$) is less than the total water potential of adjacent deeper layers (Gardner and Ehlig 1962). Thus, the sign of q in Eq. 2 above can be either positive or negative, depending upon the direction of the potential gradient. Or, stated in qualitative terms, when deeper layers are significantly wetter than the surface soil layers, water may move upward by capillary flow from wetter into dryer layers. This is particularly likely when the permanent water table is close to the soil surface.

Vapor-phase transport of significant quantities of water has also been reported (Hanks 1958; Jackson et al. 1973, 1974). The requirement here is unsaturation (sufficient air-filled porosity) and a temperature gradient such that a fugacity (Gibbs Free Energy) gradient exists. Diurnal condensation of water droplets was noted in cool, dry soil near the surface at night (Fiscus and Huck 1972), whenever moisture could be resupplied from warmer, wet soil in deeper regions of the soil profile. Photographic records of this process were made in studies at the Auburn rhizotron (Huck, Klepper, and Taylor 1970).

Under dryland conditions, soil water content is lost from surface soil layers to evaporation between rainfall events, forming an effective "mulch." As the surface layers dry, the unsaturated hydraulic conductivity of these layers is reduced to such an extent that water stored in deeper layers moves upward very slowly, retarding evaporative loss of the remaining soil water stored beneath the surface (Van Bavel, Brust, and Stirk 1968). Although often ignored, both water movement and root growth in soil cracks and worm holes can be significant (Wang, Hesketh, and Woolley 1986; Hasegawa and Sato 1987). In the absence of living root systems, the main avenues for loss of stored

water from the soil are evaporation from the surface and
drainage from deeper layers into the ground water or underground aquifers.

SOIL-ROOT-WATER INTERACTIONS

Water held in soil storage below the saturation water
content moves relatively slowly because of the exponential
relationship between soil water content and unsaturated
hydraulic conductivity (Reid and Huck, 1990). Therefore,
it forms a stable reservoir to meet between rainfall events
the water needs of growing plants. The roots of growing
plants penetrate these moist soil layers and remove water
to meet the transpiration requirement of their leaves
(Moreshet and Huck, 1990). If the root system cannot
replace transpirational water losses, the plant will not
thrive. When drought stress becomes more severe, longer
term growth and metabolic processes are affected adversely
(Huck et al. 1986).

Many adaptive strategies have evolved to facilitate
root water uptake. New root growth occurs predominantly in
moist soil where the hydraulic resistance to unsaturated
flow is much lower than in dryer soils. Once inside the
root system, specialized conducting cells of the xylem tissue facilitate long-distance transport from moist soil to
the transpiring leaves (Reid and Hutchison 1986). The rate
of new root growth in moist soil can be significantly
greater than the rate of unsaturated water flow. Roots,
therefore, may profit by growing into deeper layers where
water is more freely available rather than waiting for
water to move into them by capillary flow, despite the
additional carbon required for new root growth (see model
description below; Huck and Hillel 1983; Hoogenboom and
Huck 1986).

Soil layers with significant numbers of living roots
lose water much more rapidly than would be expected if the
only method of water loss from those layers were unsaturated flow (Van Bavel, Stirk, and Brust 1968; Allmaras,
Nelson, and Voorhees 1975). In general, water uptake is
directly proportional to root length density (Taylor and
Klepper 1975). A longer root system, however, will not
always lead to increased total water uptake. Eavis and
Taylor (1979) found an inverse relationship between total
root length and water uptake per unit of root under uniform
moisture conditions.

Root Growth

The spatial distribution of active roots changes over the course of a growing season. Root tips extend into deeper, wetter soil regions to meet the plant's continuing need for water. The location of available soil water changes in response to rainfall, drainage, and evaporation processes. Between rainfall or irrigation events, most of the new root expansion occurs in deeper soil layers, where conditions for root growth are more favorable (Taylor and Klepper 1978) and where stored water is available for uptake (Mayaki, Teare, and Stone 1976; Willatt and Taylor 1978; Garay and Wilhelm 1983). When the surface layers are rewet by subsequent rainfall, root growth in the deeper soil layers declines, and another generation of new roots begins growth in surface layers, which generally contain more minerals and oxygen. Thus root growth in the surface is favored so long as adequate water is available to support the growth.

The life span of an individual small feeder-root usually extends from a few days to a few weeks (Huck 1977; Huck, Hoogenboom, and Peterson 1987). Thus many roots that developed in surface soil during early stages of development begin to die as the surface dries within a week or two after emergence of a typical annual plant. Because of the continuing regrowth of active roots, with new root tips extending into soil regions with favorable conditions while older roots die back in other soil regions, the spatial distribution of an aggregate root system varies markedly over the course of a growing season.

The physical mechanism that controls new root growth is not well understood. Lang and Thorpe (1986) used radiotracers to show that osmotic potential of the water surrounding a growing root can influence biomass partitioning, and thus the pattern of root growth in a nonhomogeneous soil. In addition to water availability, root growth rate is a function of mechanical resistance to deformation of the soil (Greacen and Oh 1972) and to the availability of adequate aeration in the soil pore spaces (Huck 1968, 1977).

Root Distribution

Plants with a deep rooting system often produce more above-ground dry matter than those with shallow root systems caused by unfavorable soil conditions such as a compacted plow layer (Miller 1987) or acidic subsoil condition (Taylor, Huck, and Klepper 1972). This occurs because of the larger amount of soil water that is potentially available to the deeper rooted plant. A deeper rooting system does not always guarantee maximum yield, however; if soil water and nutrients are available in ample quantity, a relatively small root system may be sufficient to support a large canopy and to sustain a high yield (Taylor 1983; Van Noordwijk and de Willigen 1987). Sometimes, a crop actually produces more roots than required for maintaining optimal growth (McGowan and Tzimas 1985).

The distribution of living roots within the soil profile varies in both space and time (Willatt and Taylor 1978; Proffitt, Berliner, and Oosterhuis 1985; Hoogenboom, Huck, and Hillel 1987). New root growth depends upon the activity of the above-ground parts of the plants and thus is sensitive to both biological and climatological variables. During daylight, the leaves form sugars through the photosynthetic process. A portion of these sugars is utilized by the root system to fuel the metabolism of roots already present and to form the structural components of additional or new roots.

Availability of stored soil water is one of the most critical factors influencing new root growth and thus the distribution of active roots in the soil profile. Under high-rainfall or irrigated conditions, most of the new roots will be concentrated near the soil surface, which will be at or near field capacity (Blum and Arkin 1984; Huck et al. 1986). Under dryland conditions or during long periods of drought, most of the active roots will be found in the deeper portions of the soil profile because of the relatively higher water contents of those layers compared with the upper layers of the soil profile (Proffitt, Berliner, and Oosterhuis 1985; Hoogenboom, Huck, and Hillel 1987).

WATER UPTAKE BY ROOTS

Plants have both active and passive mechanisms for extracting water from soil (Weatherley 1975, 1982; Kramer 1983). Osmotic or active absorption occurs in response to solute accumulation in the root, when the osmotic potential of the root tissue is below that of ambient soil water (Brieger 1928; Broyer 1947). Water then diffuses from the soil into the root system. This phenomenon can often be observed experimentally as guttation effects in detopped plants growing on a well-watered (moist and aerated) soil.

More often, however, water uptake is a passive process (Weatherley 1975, 1982). As the canopy transpiration rate increases during daylight, water potential gradients between shoot and root tissue become steeper, thereby increasing the rate of water flow through the connecting xylem elements (Huck 1985b). The water is actually "pulled" upwards through cohesion of the water molecules (Pickard 1981; Nobel 1983). Through the process of transpiration, water vapor is lost through open stomata. Water vapor evaporating from inside the sub-stomatal cavities is replenished by new molecules of water supplied by xylem vessels connected with other parts of the plant. The whole catenary process causes water to move through interconnecting xylem vessels toward the leaves, which are at a lower water potential than the roots and surrounding soil.

Uptake Mechanisms

As root water potential drops because of xylem water movement into the shoot, the water potential gradient between roots and the surrounding soil is maintained, allowing for continued root water uptake, even under drought conditions (Oosterhuis 1987). There is a linear relationship between the total transpiration rate and the pressure potential difference in both barley and lupin (Passioura 1984; Passioura and Munns 1984; Munns and Passioura 1984). In general, root water uptake rates closely follow transpiration rates, so that an increase in the transpiration rate quickly results in an increase in root water uptake (Hirasawa, Araki, and Ishihara 1987) and, in some cases an apparent reduction in root resistance (Stoker and Weatherley 1971, Reid and Huck, 1990). There is, however, a theoretical maximum root water uptake rate per unit root length that is not affected by either the potential difference between roots and the soil or by other differences in relative water content (Lang and Gardner 1970).

When the aggregate water uptake rate (the combined activity of the entire root system) lags the aggregate transpiration rate (from all leaves), the plant's overall water content decreases. Blackman and Davies (1985) postulated that a signal sent from the root system to the shoot, during the initial stages of soil drying (a result of root water uptake), induces partial stomatal closure in the leaves. As leaf tissue water content is further reduced in the onset of drought stress, stomatal closure and a reduction in both transpiration and carbon dioxide exchange rates (apparent net photosynthesis) occur (Huck et al. 1983; Cox and Joliff 1987). Plants also have adaptive processes to reduce leaf water potential as soil water potential decreases (Dirksen 1985).

Resistance to Water Movement

The movement of water from the soil into the roots and into the shoot involves several components, each of which may be characterized by a specific resistance to water movement (Moreshet and Huck, 1990). The following components are found in the rhizosphere: soil resistance, R_{soil}; soil-root interface resistance, $R_{interface}$; and root resistance, R_{root}. R_{root} has a radial component, R_{radial}, and an axial component, R_{axial} (Gardner 1960; Newman 1969; Boyer 1971; Faiz and Weatherley 1977). There is some controversy in the literature concerning which of these component resistances is the most critical factor in limiting water uptake rates (e.g., Tinker 1976; Passioura 1980, 1981; Oosterhuis 1983). Depending on the soil moisture conditions, any of these resistances could limit the effective water uptake rate. Under very wet conditions R_{radial} is limiting, while under very dry conditions R_{soil} may be the rate-limiting resistance.

The process of plant water uptake from storage in the bulk soil begins with capillary movement toward the root surface. In most cases, this process can be characterized as unsaturated flow in either the horizontal or vertical direction (or some of each; Arya, Blake, and Farrell 1975a, 1975b; Arya, Farrell, and Blake 1975) across a perirhizal gradient (Faiz and Weatherley 1978). Whenever water is removed from soil storage by the action of roots, the soil water potential in the immediate vicinity is reduced to slightly below that of the surrounding bulk soil (Faiz and Weatherley 1978; Zur et al. 1982; McCoy et al. 1984). This creates a potential gradient in the direction of the roots, which in turn causes water to flow toward them (Molz and Klepper 1972; Molz 1975, 1976). In general, R_{soil} increases with a decrease in soil hydraulic conductivity and soil water potential (Allmaras, Nelson, and Voorhees 1975), so the gradient must continue to increase as water is removed from storage in the soil.

The second step in the process is water movement across the soil-root interface. In moist soils, the con-

tact between the root surface and adjacent soil particles
is generally very good; as soil surrounding the root surface dries from root water uptake or surface evaporation,
it may shrink away from the root. Water is replaced by
air, and the perirhizal potential gradient increases significantly (Faiz and Weatherley 1978). As a result, the apparent resistance to flow across the soil-root interface increases greatly (Weatherley 1976; Moreshet et al. 1987).
Diurnal changes in root-soil gaps were measured by Huck,
Klepper, and Taylor (1970). Faiz and Weatherley (1982)
have shown experimental evidence for the importance of this
interface, but quantitative measurements of its magnitude
are not available at present. Field studies with potted
trees (Legge 1985) showed no significant effects of
$R_{interface}$ on water uptake.

New techniques have recently been introduced from the
medical disciplines, such as using Computer-Assisted Tomography (CAT) to estimate water flow rates (Hainsworth and
Aylmore 1983; Brown et al. 1987) and water extraction by
roots (Hainsworth and Aylmore 1986, Rogers and Bottomley
1987). In some cases, these studies have shown that assumptions related to the resistances involved in the water uptake process were inadequate, and that further refinements
of theory are required (Hainsworth and Aylmore 1986), especially when conclusions are based on simplified models
(Taylor, Klepper, and Rickman 1979; Passioura 1984).

Once water passes from the soil and through the root
epidermis, the remaining transport processes occur inside
the plant (Johnson, Brown, and Kramer 1987). The main bar-
rier against free movement of water within the root tissue
is the endodermal layer, or the so-called Casparian strips,
which constrain water to pass through protoplasm as it
enters the xylem elements of the stele. These endodermal
flow barriers effectively increase root resistance to water
uptake, but they may also reduce the flow of water from the
plant back into the soil. Water movement from living roots
into the surrounding soil has been measured (Molz and Peter
son 1976; Van Bavel and Baker 1985), but a high $R_{interface}$
between roots partially dried soil material probably re-
stricts the maximum rate of water release from the roots
back into the soil (Dirksen and Raats 1985).

After passing through the endodermis, water enters the
xylem and moves longitudinally through the various secondary roots toward the primary root. Root xylem elements are
connected to homologous components of the stem through a
"transition zone" between the roots and stem; once in the
stem xylem, water is distributed to leaves, branches,

stems, and fruits. The axial resistance of root xylem is usually much lower than the other resistances involved in water uptake (Klepper 1983). Fowkes and Landsberg (1981) found that the number of xylem vessels in apple trees is directly proportional to the root radius, and that the R_{axial} is inversely proportional to the root radius. Although the direction of water flow in the xylem is generally upward, Kirkham (1983) showed that water flow could occur in either direction depending upon environmental conditions.

Most of the water uptake by root systems occurs in the younger root tissues (Clarkson and Robards 1975), where suberization and lignification processes are incomplete. The R_{radial} of young root tissue is much lower than the R_{radial} of suberized roots. In general, the path of water through the soil-root system will be that of least resistance. Water uptake per unit root length is affected by the total root length density and the spacing and distribution of the roots within a soil horizon (Landsberg and Fowkes 1978) as well as by specific resistances within and directly surrounding the root.

Root hairs may also play an important role in the process of water and solute uptake (Weatherley 1982). Because root hairs are so small, they can be easily overlooked. Root hairs increase the soil-root interface area and therefore potentially reduce the resistance to water uptake. They are formed on the root epidermis of many species, originating from specialized cells called trichoblasts. Root hairs are not found in either very wet or very dry conditions (Kramer 1983). Huck (1968) reported that root hair formation in tomato (Lycopersicon esculentum) was strongly dependent upon the availability of molecular oxygen at the root surface.

Ectomycorrhizal mycelia are also involved in the water uptake process by roots. Experiments with trees have shown that water can flow through mycorrhizae hyphae to seedling roots existing in soils with a high moisture content (Brownlee et al. 1983; Ruehle 1983). Mycorrhizae serve to increase the effective root surface area, thereby reducing the root-soil interface resistance, particularly under relatively dry soil conditions (Reid 1985).

ROOT AND SOIL WATER MODELING

Interactions

Soil water availability influences both root growth rates and the spatial distribution of new root growth. At the same time, the number and location of active roots in a given soil volume exerts a marked influence upon water uptake patterns, and hence the spatial distribution of water - which in turn affects future root growth rates. This system behavior is very complex, and includes many feed-back and feed-forward loops. To better understand the complexity of these systems, detailed computer simulation models that describe the interaction of all relevant processes in the plant-soil-atmosphere continuum are needed (Hoogenboom, Huck, and Hillel 1987; Schulze et al. 1987; Grant et al. 1989). Such models are useful for enhancing our understanding of system behavior, and for studying which sections of the system may be most sensitive to environmental changes or biological adaptations. Unfortunately, many of the models that deal with water transport are simplified, mainly because of the complexity of the system and difficulties in measuring the parameters involved (Molz 1981). Many plant water uptake models, for example, have no active root growth component.

A recent study by Hoogenboom, Huck, and Peterson (1988) considered both shoot and root systems and their interactions, with the aid of a whole-plant computer simulation model, called ROOTSIMU. In companion experiments, root and shoot growth of soybean plants were studied at the Auburn rhizotron (Huck and Taylor 1982) under rain fed and irrigated conditions (Huck et al. 1983). One of the main objectives of this project was to investigate plant responses to environmental stresses (e.g., drought), comparing performance of stressed plants with that of controls that were irrigated whenever soil water potential in the root zone fell below -10 centibars. Soybean (Glycine max. [L.] Merr.) was used in this project because of its importance as a food crop.

The results of these studies (Huck et al. 1983; Hoogenboom, Huck, and Peterson 1986, 1987; Hoogenboom, Peterson, and Huck 1987) were used to formulate differential equations, which were then combined into a computer model to simulate plant growth and function in a real-world environment (Huck and Hillel 1983; Hoogenboom and Huck 1986; Hoogenboom, Peterson, and Huck 1987, Hoogenboom, Huck, and

Peterson 1988). Special attention was given to the relationship between the shoot and root systems, and to the
interactions that occur as a result of stress conditions
imposed upon growing plants. The system boundaries of the
model used in these studies (Hoogenboom and Huck 1986) are
illustrated in Figure 13.3, which shows a hypothetical soybean plant as it is considered by the model. The soil profile is treated as a set of horizontal layers, with root
growth and death rates, water uptake, and soil water content calculated at each time-step for each soil layer.

The water balance flow diagram shown as a Forrester
diagram in Figure 13.2 involves some simplifications. Numerical calculations in the plant and soil water balance
sections of the model consider flows indicated by arrows in
the Figure. Water from rainfall or irrigation enters the
system by infiltration into the soil profile through the
surface layer, and then penetrates into deeper soil horizons by saturated flow. It can leave the system in several
ways: by drainage from the bottom of the profile, by evaporation from the soil surface, or by root water uptake and
subsequent transpiration from the leaves. Calculation of
an internal water balance for the plant (right side of
Figure 13.2) permits an estimate of plant water potential,
which drives the water uptake process through equations
similar to those given earlier in this review.

An overview of the carbon balance, illustrating relationships between the different vegetative organs of the
plant, is presented in Figure 13.4. The model accounts for
growth processes, relating carbon dioxide fixation (photosynthesis), carbohydrate partitioning, and metabolism, as
well as death of root, stem, and leaf tissue to keep a continuous accounting of active plant biomass. In the model,
photosynthesis (net carbon fixation rate) is represented as
a function of incoming total energy (solar radiation), carbon dioxide concentration of the air, and the total active
leaf area of the plant (LAI or leaf area index).

As the LAI increases, evaporative demand (at a given
level of solar insolation) also increases. This increased
demand must be met by water supplied from the root system.
Carbohydrates produced through the photosynthetic process
are stored in the soluble carbohydrate pool shown near the
top of Figure 13.4. A proportion of this pool is used for
maintenance respiration by the shoot and root system, while
the remainder is used for growth of new shoot or the root
tissue.

Partitioning between the shoot and root systems is
controlled by the level of water stress (Huck and Hillel

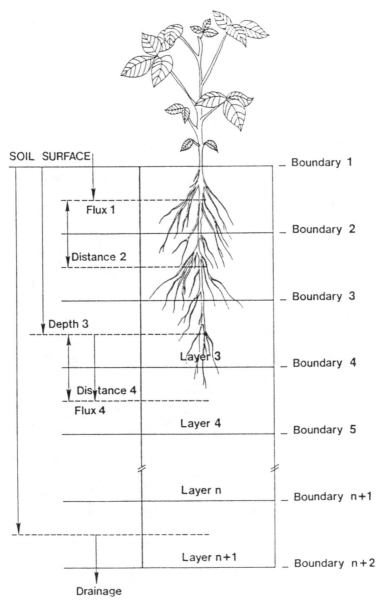

Figure 13.3. System boundaries for water movement in a soil with an active root system (hypothetical soybean plant growing in a one-dimensional layered soil consisting of infinite uniform layers in the horizontal direction, after Hoogenboom and Huck 1986).

288

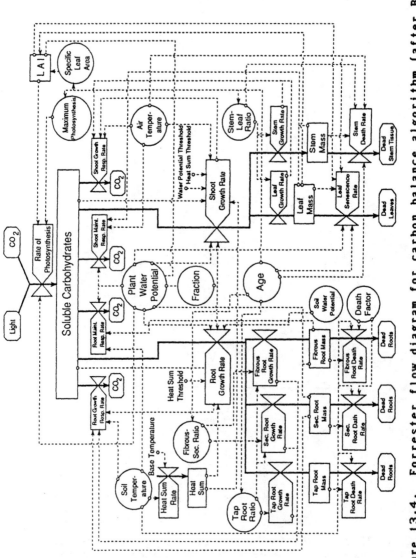

Figure 13.4. Forrester flow diagram for carbon balance algorithm (after Blake and Hoogenboom 1988).

1983), with most of the carbohydrates going to support shoot growth under well-watered conditions and an increasing fraction allocated for root growth as drought stress begins to develop in the simulated plant (low plant water potential, resulting from evaporative demand exceeding root uptake rate). Stomatal closure, which occurs at low leaf water potential, reduces both transpiration and net photosynthetic rates. This allows root uptake to catch up with transpiration demand preventing severe dessication of leaf tissues, but it also reduces carbon fixation rates and thus total plant growth.

An example of simulated root and shoot growth rates is shown in Figure 13.5 for an irrigated versus a nonirrigated crop during a ten-day period (Hoogenboom, Huck, and Peterson 1988). Shoot growth of the irrigated plants exceeded that of nonirrigated plants throughout the period shown. Although the actual amount of root growth for the irrigated plants exceeded that of the rain-fed plants, the fraction of available carbohydrates which the irrigated plants invested in root growth was substantially lower. (The value of FRAC shown in the Figure 13.5 represents the fraction of total available carbohydrates allocated to shoot growth, so whenever FRAC increases, the remaining carbohydrates which can be allocated for root growth must be reduced.) Most of the simulated shoot and root growth occurred during afternoon and early evening hours, when carbohydrate levels approached daily maximum values, resulting from an excess of photosynthetic rates over combined tissue respiration rates. Peak root growth rates are slightly lower on the last two days of this ten-day simulation period because light rains on day 205 and 208 reduced drought stress in both treatments on those days (Hoogenboom, Huck, and Peterson 1988).

Root Growth

Beyond the initial partitioning of carbohydrates between root and shoot systems, there must be further allocation among the roots growing at different depths. The model allocates root growth according to the amount of tissue already present in a given soil layer and the soil water availability in that layer. The model considers the root growth rate for any given soil layer to be a function of root mass, temperature, carbohydrate availability, and soil water content; the root death rate is a function of root mass and temperature only. The change in root length

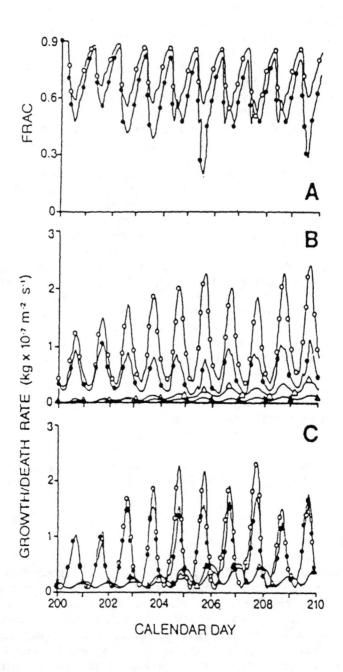

Figure 13.5. Simulated shoot and root growth rates for irrigated or non-irrigated plants (after Hoogenboom, Huck, and Peterson 1988).

per unit soil volume is the algebraic sum of root growth and death rates (Huck, Hoogenboom, and Peterson 1987).

When the soil profile is uniformly moist, most of the new root growth occurs in the top layers; under dry conditions, root growth occurs primarily in the deeper, wetter soil horizons. The water uptake rate for roots in each layer, and for the aggregated root system, is a function of both plant and soil water potentials. This follows from Eq. 1, which states that water flux velocity is proportional to the difference between soil and plant water potential.

The long-term effects of water stress upon simulated root growth of irrigated and nonirrigated plants are illustrated in Figure 13.6, which shows the total length of active root present (by depth and treatment) for a simulated 100-day growing period (after Hoogenboom, Huck, and Peterson 1988). Weather data from Auburn, Alabama, measured during the 1981 season, served as the driving function for this simulation. The soil profile was arbitrarily represented by 0.1-, 0.15-, and 0.20-m thick layers, down to a depth of 1.40 m; soil profile characteristics were taken as those of a Marvyn loamy sand (fine-loamy, siliceous, thermic, Plinthic Paleudult), which was used in companion validation experiments.

These simulation studies showed that during the first twenty-five days of the growing period, most of the increase in root length occurred in the top 0.6 m of the soil profile. Some time was required for roots to penetrate into the deeper soil layers. The model predicted a marked increase in root length below 0.6 m in both treatments between calendar days 180 and 210, when a drought period occurred and no rainfall was observed. Heavy rains were measured between calendar days 220 and 230; as a result, the upper layers were rewet. Simulated root growth declined in the lower soil layers and increased in the upper layers.

The model performance, in predicting spatial distribution of active roots over the course of a growing season and accounting for variation in soil water content, was qualitatively in agreement with experimentally measured data. Simulated net root growth rates (algebraic sum of root growth and root death rates) are shown in Figure 13.7 (after Hoogenboom, Huck, and Peterson 1988; note difference in time scales between Figure 13.6 and Figure 13.7).

At the beginning of the ten-day period, the simulated irrigated plants formed more roots than the nonirrigated plants, mainly between soil depths of 1.0 and 1.2 m. Sim-

292

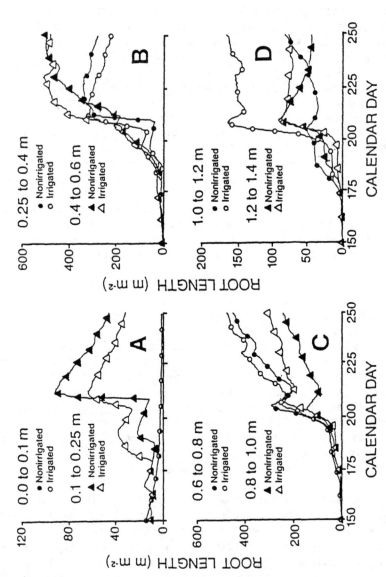

Figure 13.6. Simulated root length from calendar day 150 to 250 for depths indicated (after Hoogenboom, Huck, and Peterson 1988).

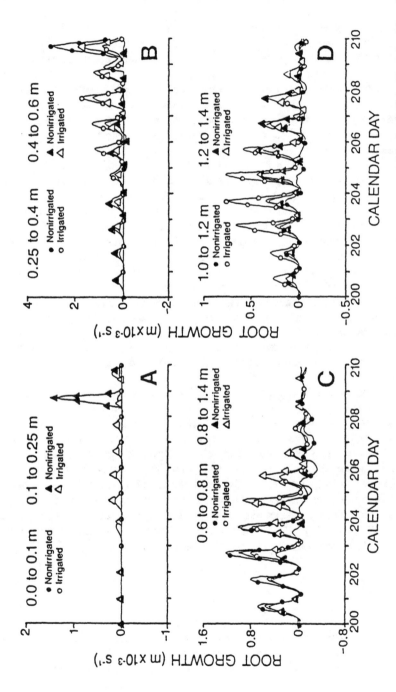

Figure 13.7. Simulated root growth rates from calendar day 200 to 210 for depths indicated (after Hoogenboom, Huck, and Peterson 1988).

ulated nonirrigated plants formed more roots between depths of 0.6 and 0.8 m. The model did not predict any root growth in the top layer because most of the available soil water had been lost to surface evaporation, and therefore the soil was too dry to support any new root growth. As the drought period continued, water was applied to the irrigated treatment, and an increase in root growth between a depth of 0.1 and 0.4 m was predicted soon after the top layers of the soil profile were rewet. At the same time a net decrease in root length was predicted in the lower layers of the irrigated soil.

After a rain on calendar day 208, nonirrigated plants initiated new root growth between depths of 0.25 and 0.6 m, while root growth slowed at the 0.6 to 1.4 m depth. Simulated death continued in the irrigated treatment between a depth of 0.8 and 1.4 m, causing a net reduction in root length in those soil layers. The irrigated plants showed no significant increase in simulated root growth following the rainfall events. Instead, the model predicted that increased water in the soil (which was already wet from irrigation) would reduce resistance to water movement, resulting in an increase in the rate of water uptake and total plant water content. This, in turn, reduced the fraction of soluble carbohydrates partitioned into the root system and further reduced total simulated root growth after the rain.

The net effect of the root growth processes illustrated in Figure 13.7 can be seen by reference to Figure 13.8, in which the vertical root profiles are shown for the same time period. Very little new root growth occurred below 1 m between day 205 and day 210. Following the rain on day 203 (Figure 13.9), nearly all new root growth was in the shallower soil layers. By contrast, the very large number of new roots which grew below 1 m depth in the nonirrigated treatment (which grew in response to the water stress from day 190 onwards) died back quickly after day 205 when water was again available closer to the soil surface.

Soil Water

Soil water potential in each soil layer changes continuously because of the interactions of many processes in the soil-root system. Water moves through the soil layers, either by saturated or unsaturated flow depending on the amount already present in the different soil layers. The timing and amounts of experimentally measured rainfall

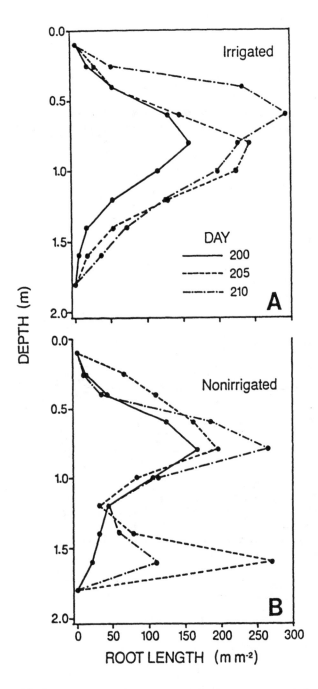

Figure 13.8. Vertical root distribution profile simulated for days 200, 205, and 210 (after Hoogenboom, Huck, and Peterson 1988).

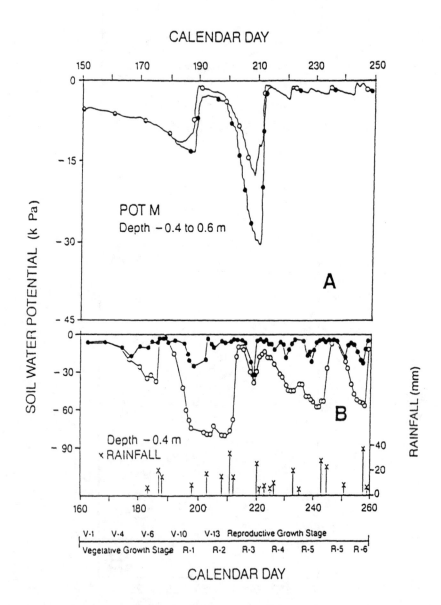

Figure 13.9. Simulated soil water potential at a depth of 0.4 m (above), compared with tensiometer measurement and rainfall data (below), as reported by Hoogenboom, Huck, and Peterson (1988).

(Hoogenboom, Huck, and Peterson 1988) and the soil water potential for each treatment as measured by tensiometers are shown in Figure 13.9B. For comparison, the simulated soil water potential for profile depths between 0.4 and 0.6 m for the same time period is shown in Figure 13.9A. The simulated results show a decrease in soil water potential from calendar day 170 to 185 because of a ten-day drought period. A similar decrease is predicted from calendar day 195 to 210, when soil evaporation and transpiration exceeded precipitation, and the overall soil profile showed a net loss of water. Several heavy rains between day 210 and 225 resulted in large increases in soil water potential for both treatments, with the irrigated profile reaching saturation much earlier.

Simulated soil water potential is shown as a time-series function for only a single layer (corresponding to the depth of tensiometers in the measurement data set) in Figure 13.9. The vertical distribution of soil water for three representative days is shown for calendar days 200, 205, and 210 in Figure 13.10. There are marked differences in soil water distribution between irrigated (Figure 13.10A) and nonirrigated (Figure 13.10B) profiles. In general, the irrigated profile has a higher soil water content for all depths than the nonirrigated profile. During the ten-day period shown, soil water content decreased in most layers because transpiration and evaporation continued whether or not there was a rain.

Instantaneous root water uptake rates (computed as potential gradient between leaf and soil water potential for each layer divided by the sum of applicable soil and plant resistances) are shown for this same ten-day period in Figure 13.11). No water extraction was predicted from the top soil layer because the surface soil was very dry. The model also predicted no root growth in these layers (Figure 13.7). The simulated irrigated plants showed an increasing water uptake rate from the second layer (0.10 to 0.25 m depth) as irrigation water began to infiltrate into this layer on calendar day 204. The marked increase in water uptake from the same soil layer by roots of the non-irrigated plants on calendar day 209 resulted from the greatly increased soil water content at that depth following a heavy rain on calendar day 208.

In the deeper layers, simulated water uptake rate for the irrigated plants followed the same diurnal curve for each day, with most of the water being extracted from a depth of 0.4 to 0.8 m. Simulated nonirrigated plants extracted most of their water from the deeper soil layers.

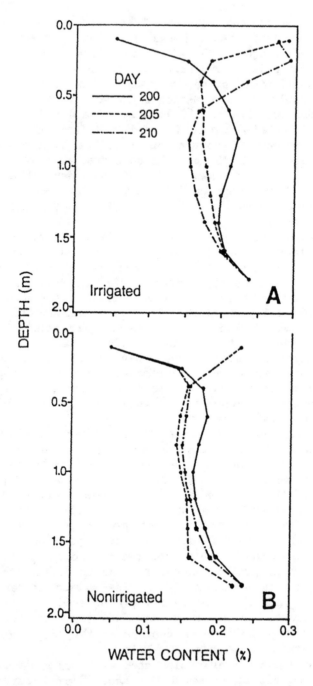

Figure 13.10. Vertical water content profile simulated for days 200, 205, and 210 (after Hoogenboom, Huck, and Peterson 1988).

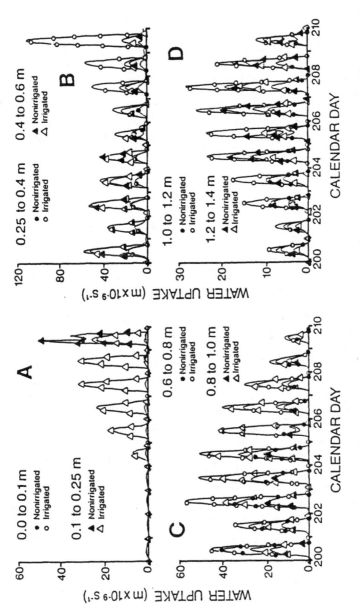

Figure 13.11. Simulated water uptake rates from calendar day 200 to 210 for roots growing at the depth and treatment indicated (after Hoogenboom, Huck, and Peterson 1988).

Partial stomatal closure, predicted in the nonirrigated treatment as a result of drought stress, resulted in lower soil water extraction rates compared with water uptake of the irrigated control plants.

Total accumulated water uptake as a function of depth is shown at ten-day increments for the entire growing season in Figure 13.12. Most of the simulated water uptake from the irrigated soil profile occurred in the upper half of the profile, where most of the root growth was also predicted. For the nonirrigated treatment, the model predicted more roots growing in the deeper soil layers. Since the surface layers of the non-irrigated treatment were also drier, the proportion of water extracted from subsoil layers by non-irrigated root systems gradually increased as the growing season progressed.

SUMMARY

Plants obtain needed water and minerals from the soil in which they grow by absorption through their root systems. Water is lost from the soil surface by evaporation or removed from the deeper regions of the soil profile by internal drainage and the activity of living roots.

Growth of new roots within the soil profile is very sensitive to availability of water. As soil water content decreases in the upper portions of the soil profile due to evaporation and water extraction, roots grow into the deeper, wetter soil regions where water is available for uptake. Following heavy rain or irrigation, root growth in deeper soil regions declines, and shallower roots are regenerated. Roots can take up water from the soil environment whenever the root xylem water potential is below that of water stored in adjacent regions of the soil profile.

Complex systems of this nature can be described by models. Water movement through both soil and plant tissues is described in a dynamic computer simulation model. Examples of output from simulation runs with the model predicted that most of the water extracted by a soybean root system would come from those soil regions having the highest water content. As the soil water content decreased in the upper layers, roots grew into the deeper soil layers, where additional water was available at relatively high water potential and where the combined flow resistance terms were less than in the drier layers above. Whenever the soil was rewet through rainfall or irrigation, soil water content increased initially in the top layers. Soil water percolat-

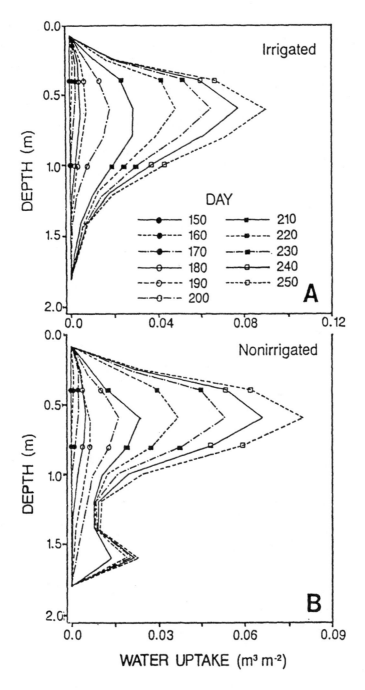

Figure 13.12.   Vertical water uptake profiles computed at ten-day intervals from day 150 to day 250 for each treatment (after Hoogenboom, Huck, and Peterson 1988.)

ed downward through the profile and gradually the deeper layers began showing increases in soil moisture content. New root growth followed changes in soil water content, and the cycle of new root growth, followed by death, was repeated on successive irrigation cycles.

REFERENCES

Allmaras, R. R., W. W. Nelson, and W. B. Voorhees. 1975a. Soybean and corn rooting in southwestern Minnesota: I. Water-uptake sink. Soil Science Society of America Proceedings 39:764-771.

Allmaras, R. R., W. W. Nelson, and W. B. Voorhees. 1975b. Soybean and corn rooting in southwestern Minnesota: II. Root distributions and related water inflow. Soil Science Society of America Proceedings 39:771-777.

Arya, L. M., D. A. Farrell, and G. R. Blake. 1975. A field study of soil water depletion patterns in presence of growing soybean roots: I. Determination of hydraulic properties of the soil. Soil Science Society of America Proceedings 39:424-430.

Arya, L. M., G. R. Blake, and D. A. Farrell. 1975a. A field study of soil water depletion patterns in presence of growing soybean roots: II. Effect of plant growth on soil water pressure and water loss patterns. Soil Science Society of America Proceedings 39:430-436.

Arya, L. M., G. R. Blake, and D. A. Farrell. 1975b. A field study of soil water depletion patterns in presence of growing soybean roots: III. Rooting characteristics and root extractions. Soil Science Society of America Proceedings 39:437-444.

Baer, J. 1969. Dynamics of Fluids in Porous Media. Amsterdam, Netherlands: Elsevier.

Blackman, P. G., and W. J. Davies. 1985. Root to shoot communication in maize plants of the effects of soil drying. Journal of Experimental Botany 36:39-48.

Blake, J. I., and G. Hoogenboom. 1988. A dynamic simulation of loblolly pine (Pinus taeda L.) seedling establishment based upon carbon and water balances. Canadian Journal of Forest Research 18:833-850.

Blum, A., and G. F. Arkin. 1984. Sorghum root growth and water-use as affected by water supply and growth duration. Field Crops Research 9:131-142.

Boyer, J. S. 1971. Resistances to water transport in soybean, bean, and sunflower. Crop Science 11:403-407.

Brieger, F. 1928. Untersuchungen über die Wasseraufnahme ganzer Pflanzen. Jahrbuch Wissenschaftlichen Botanik 69:295-330.

Bristow, K. L., G. S. Campbell, and C. Calissendorff. 1984. The effects of texture on the resistance to water movement within the rhizosphere. Soil Science Society of America Journal 48:266-270.

Brooks, R. H., and A. T. Corey. 1964. Hydraulic properties of porous media. Hydrology Paper no. 3. Fort Collins: Colorado State Univ.

Brown, J. M., W. C. Fonteno, D. K. Cassel, and G. A. Johnson. 1987. Computed tomographic analyses of water distribution in three porous foam media. Soil Science Society of America Journal 51:1121-1125.

Brownlee, C., J. A. Duddridge, A. Malibari, and D. J. Read. 1983. The structure and function of mycelial systems of ectomycorrhizal root with special reference to their role in forming inter-plant connections and providing pathways for assimilate and water transport. Plant and Soil 71:433-443.

Broyer, T. C. 1947. The movement of materials into plants. Part I. Osmosis and the movement of water into plants. The Botanical Review XIII:1-58.

Clarkson, D. T., and A. W. Robards. 1975. The endodermis, its structural development and physiological role. In The Development and Function of Roots, ed. J. G. Torrey and D. T. Clarkson, pp. 415-436. London: Academic Press.

Corey, A. T., R. O. Slatyer, and W. D. Kemper. 1967. Comparative terminologies for water in the soil-plant-atmosphere system. In Irrigation of agricultural lands, ed. R. M. Hagan, H. R. Haise, and T. W. Edminster, 427-445. ASA Monograph 11. Madison, WI: American Society of Agronomy.

Craig, R. F. 1983. Soil mechanics. Wokingham, England: Van Nostrand Reinhold.

De Jong, R., C. A. Campbell, and W. Nicholaichuk. 1983. Water retention equations and their relationship to soil organic matter and particle size distribution for disturbed samples. Canadian Journal of Soil Science 63:291-302.

Dirksen, C. 1985. Relationship between root uptake-weighted mean soil water salinity and total leaf water potentials of alfalfa. Irrigation Science 6:39-50.

Dirksen, C., and P. A. C. Raats. 1985. Water uptake and release by alfalfa roots. Agronomy Journal 77:621-626.

Eavis, B. W., and H. M. Taylor. 1979. Transpiration of soybeans as related to leaf area, root length, and soil water content. Agronomy Journal 71:441-445.

Faiz, S. M. A., and P. E. Weatherley. 1977. The location of the resistance to water movement in the soil supplying roots of transpiring plants. New Phytologist 78:337-347.

Faiz, S. M. A., and P. E. Weatherley. 1978. Further investigations into the location and magnitude of the hydraulic resistances in the soil:plant system. New Phytologist 81:19-28.

Faiz, S. M. A., and P. E. Weatherley. 1982. Root contraction in transpiring plants. New Phytologist 92:333-343.

Fiscus, E. L., and M. G. Huck. 1972. Diurnal fluctuations in soil water potential. Plant and Soil 37:197-202.

Fowkes, N. D., and J. J. Landsberg. 1981. Optimal root ystems in terms of water uptake and movement. In mathematics and plant physiology, vol. 16 of Experimental botany, ed. C. W. Rose and D. A. Charlesdwards, 109-125. New York: Academic Press.

Garay, A. F., and W. W. Wilhelm. 1983. Root system characteristics of two soybean isolines undergoing water stress conditions. Agronomy Journal 75:973-977.

Gardner, W. R. 1960. Dynamic aspects of water availability to plants. Soil Science 89:63-73.

Gardner, W. R., and C. F. Ehlig. 1962. Some observations on the movement of water to plant roots. Agronomy Journal 54:453-456.

Goss, M. J., W. Ehlers, F. R. Boone, I. White, and K. R. Howse. 1984. Effects of soil management practice on soil physical conditions affecting root growth. Journal of Agricultural Engineering Research 30:131-140.

Grant, R. F., J. R. Frederick, J. D. Hesketh, and M. G. Huck. 1989. Simulation of morphological development of maize under contrasting water regimes. Canadian Journal of Plant Science 69:401-418.

Greacen, E. L., and J. S. Oh. 1972. Physics of root growth. Nature (London) New Biology 235:24-25.

Gupta, S. C., and W. E. Larson. 1979a. A model for predicting packing density of soils using particle-size distribution. Soil Science Society of America Journal 43:758764.

Gupta, S. C., and W. E. Larson. 1979b. Estimating soil water retention characteristics from particle size distribution, organic matter percent, and bulk density. Water Resources Research 15:1633-1635.

Hainsworth, J. M., and L. A. G. Aylmore. 1983. The use of computer assisted tomography to determine spatial distribution of soil water content. Australian Journal of Soil Research 21:435-443.

Hainsworth, J. M., and L. A. G. Aylmore. 1986. Water extraction by single plant roots. Soil Science Society of

America Journal 50:841-848.

Hamblin, A. P. 1985. The influence of soil structure on water movement, crop root growth, and water uptake. Advances in Agronomy 38:95-158.

Hanks, R. J. 1958. Water vapor transfer in dry soil. Soil Science Society Proceedings 22:372-374.

Hasegawa, S., and T. Sato. 1987. Water uptake by roots in cracks and water movement in clayey subsoil. Soil Science 143:381-386.

Hillel, D. 1980. Fundamentals of soil physics. New York: Academic Press.

Hirasawa, T., T. Araki, and K. Ishihara. 1987. The relationship between water uptake and transpiration rates in rice plants. Japanese Journal of Crop Science 56:38-43. (In Japanese.)

Hoogenboom, G., and M. G. Huck. 1986. ROOTSIMU V4.0. A dynamic simulation of root growth, water uptake, and biomass partitioning in a soil-plant-atmosphere continuum: update and documentation. Agronomy and Soils Departmental Series no. 109, Alabama Agricultural Experiment Station, Auburn University.

Hoogenboom, G., M. G. Huck, and D. Hillel. 1987. Modification and testing of a model simulating root and shoot growth as related to soil water dynamics. In Advances in irrigation, vol. 4, ed. D. Hillel, 331-387. Orlando, FL: Academic Press.

Hoogenboom, G., M. G. Huck, and C. M. Peterson. 1986. Measured and simulated drought stress effects on daily shoot and root growth rates of soybean. Netherlands Journal of Agricultural Science 34:497-500.

Hoogenboom, G., M. G. Huck, and C. M. Peterson. 1987. Root growth rate of soybean as affected by water stress. Agronomy Journal 79:607-614.

Hoogenboom, G., M. G. Huck, and C. M. Peterson. 1988. Predicting root growth and water uptake under different soil water regimes. Agricultural Systems 26:263-290.

Hoogenboom, G., C. M. Peterson, and M. G. Huck. 1987. Shoot growth rates of soybean as affected by drought stress. Agronomy Journal 79:598-607.

Huck, M. G. 1968. Metabolic changes in oxygen-deficient tomato roots. Ph.D. dissertation, Michigan State University, East Lansing. [Dissertation Abstracts no. 68-4156, University Microfilms, Ann Arbor, MI.]

Huck, M. G. 1977. Root distribution and water uptake pattern. In The below ground ecosystem: A synthesis of plant associated processes, ed. J. K. Marshall,

215-225. Department of Range Science Series no. 26. Fort Collins: Colorado State University.

Huck, M. G. 1985. Water flux in the soil-root continuum. In *Roots, nutrient and water influx, and plant growth*, ed. S. A. Barber and D. R. Bouldin, 47-63. ASA Special Publication no. 49. Madison, WI: American Society of Agronomy.

Huck, M. G., and D. Hillel. 1983. A model of root growth and water uptake accounting for photosynthesis, respiration, transpiration, and soil hydraulics. In *Advances in Irrigation. Vol. 2*, ed. D. Hillel, 273-333. New York: Academic Press

Huck, M. G., G. Hoogenboom, and C. M. Peterson. 1987. Soybean root senescence under drought stress. In *Minirhizotron observation tubes; Methods and applications for measuring rhizosphere dynamics*, ed. H. M. Taylor, and J. Box, 109-121. ASA Special Publication no. 50. Madison, WI: American Society of Agronomy.

Huck, M. G., K. Ishihara, C. M. Peterson, and T. Ushijima. 1983. Soybean adaptation to water stress at selected stages of growth. *Plant Physiology* 73:422-427.

Huck, M. G., B. Klepper, and H. M. Taylor. 1970. Diurnal variation in root diameter. *Plant Physiology* 45:529-530.

Huck, M. G., C. M. Peterson, G. Hoogenboom, and C. D. Busch. 1986. Distribution of dry matter between shoots and roots of irrigated and nonirrigated determinate soybeans. *Agronomy Journal* 78:807-813.

Huck, M. G., and H. M. Taylor. 1982. The rhizotron as a tool for root research. *Advances in Agronomy* 35:1-35.

Jackson, R. D., B. A. Kimball, R. J. Reginato, and F. S. Nakayama. 1973. Diurnal soil-water evaporation: Time-depth-flux patterns. *Soil Science Society of America Proceedings* 37:505-509.

Jackson, R. D., R. J. Reginato, B. A. Kimball, and F. S. Nakayama. 1974. Diurnal soil-water evaporation: Comparison of measured and calculated soil-water fluxes. *Soil Science Society of America Proceedings* 38:861-866.

Johnson, G. A., J. Brown, and P. J. Kramer. 1987. Magnetic resonance microscopy of changes in water content in stems of transpiring plants. *Proceedings of the National Academy of Science USA* 84:2752-2755.

Khatibu, A. I., R. Lal, and R. K. Jans. 1984. Effects of tillage methods and mulching on erosion and physical properties of a sandy clay loam in an equatorial warm humid region. *Field Crops Research* 8:239-254.

Kirkham, M. B. 1983. Physical model of water in a split-root system. Plant and Soil 75:153-168.

Klepper, B. 1983. Managing root systems for efficient water use: Axial resistances to flow in root systems--Anatomical considerations. In Limitations to efficient water use in crop production, ed. H. M. Taylor, W. R. Jordan, and T. R. Sinclair, 115-125. Madison, WI: American Society of Agronomy.

Klute, A. 1965. Laboratory measurements of hydraulic conductivity of saturated soil. In Methods of soil analysis, Part 1, 210-221. Madison, WI: American Society of Agronomy.

Klute, A. 1982. Tillage effects on the hydraulic properties of soil: A review. In Predicting tillage effects on soil physical properties and processes, 29-43. Madison, WI: American Society of Agronomy.

Kramer, P. J. 1983. Water Relations of Plants. Orlando, FL: Academic Press.

Landsberg, J. J., and N. D. Fowkes. 1978. Water movement through plant roots. Annals of Botany 42:493-508.

Lang, A. R. G., and W. R. Gardner. 1970. Limitation to water flux from soils to plants. Agronomy Journal 62:693-695.

Lang, A., and M. R. Thorpe. 1986. Water potential, translocation, and assimilate partitioning. Journal of Experimental Botany 37:495-503.

Legge, N. J. 1985. Water movement from soil to root investigated through simultaneous measurement of soil and stem water potential in potted trees. Journal of Experimental Botany 36:1583-1589.

McCoy, E. I., L. Boersma, M. L. Ungs, and S. Akrtanakul. 1984. Toward understanding soil water uptake by plant roots. Soil Science 137:69-77.

McGowan, M., and E. Tzimas. 1985. Water relations of winter wheat: the root system, petiolar resistance and development of a root abstraction equation. Experimental Agriculture 21:377-388.

Marshall, T. J., and J. W. Holmes. 1979. Soil physics. Cambridge, England: Cambridge University Press.

Mayaki, W. C., I. D. Teare, and L. R. Stone. 1976. Top and root growth of irrigated and nonirrigated soybeans. Crop Science 16:92-94.

Miller, D. E. 1987. Effect of subsoiling and irrigation regime on dry bean production in the Pacific Northwest. Soil Science Society of America Journal 51:784-787.

Millington, R. J., and D. B. Peters. 1970. Transport in the soil-plant-atmosphere continuum. Scientia (January-February):1-27.

Molz, F. J. 1975. Potential distributions in the soil-root system. Agronomy Journal 67:726-729.

Molz, F. J. 1976. Water transport in the soil-root system: Transient analysis. Water Resources Research 12:805-807.

Molz, F. J. 1981. Models of water transport in the soil-plant system: A review. Water Resources Research 17:1245-1260.

Molz, F. J., and B. Klepper. 1972. Radial propagation of water potential in stems. Agronomy Journal 64:469-473.

Molz, F. J., and C. M. Peterson. 1976. Water transport from roots to soil. Agronomy Journal 68:901-904.

Molz, F. J., and I. Remson. 1970. Extraction term models of soil moisture use by transpiring plants. Water Resources Research 6:1346-1356.

Moreshet, S., and M. G. Huck. 1990. Dynamics of water permeability. In Plant Roots - the Hidden Half, ed. Y. Waisal, U. Kafkafi, and A. Eschel. [In Press]. Marcel Dekker: New York.

Moreshet, S., M. G. Huck, J. D. Hesketh, and D. B. Peters. 1987. Measuring hydraulic conductance of intact soybean root systems. In Proceedings of international conference on measurement of soil and plant water status, ed. R. J. Hanks and R. W. Brown, 221-229. Logan: Utah State University.

Munns, R., and J. B. Passioura. 1984. Hydraulic resistance of plants. III. Effects of NaCl in barley and lupin. Australian Journal of Plant Physiology 11:351-359.

Newman, E. I. 1969. Resistance to water flow in soil and plant. I. Soil resistance in relation to amounts of root: theoretical estimates. Journal of Applied Ecology 6:1-12.

Nobel, P. S. 1983. Biophysical plant physiology and ecology. New York: Freeman and Company.

Oosterhuis, D. M. 1983. Resistances to water flow through the soil-plant system. South African Journal of Science 79:459-465.

Oosterhuis, D. M. 1987. A technique to measure the components of root water potential using screen-caged thermocouple psychrometers. Plant and Soil 103:185-288.

Passioura, J. B. 1980. The transport of water from soil to shoot in wheat seedlings. Journal of Experimental Botany 31:333-345.

Passioura, J. B. 1981. Water collection by roots. In The physiology and biochemistry of drought resistance in plants, ed. L. G. Paleg, and D. Aspinall, 39-53. New York: Academic Press.

Passioura, J. B. 1984. Hydraulic resistance of plants. I. Constant or variable? Australian Journal of Plant Physiology 11:333-339.

Passioura, J. B., and R. Munns. 1984. Hydraulic resistance of plants. II. Effects of rooting medium, and time of day, in barley, and lupin. Australian Journal of Plant Physiology 11:341-350.

Peterson, C. M., M. G. Huck, and G. Hoogenboom. 1986. Proper timing is necessary for full benefit of irrigation by soybeans. In Highlights of Agricultural Research 33:14. Alabama Agricultural Experiment Station, Auburn University.

Philip, J. R. 1969. Hydrostatics and hydrodynamics in swelling soils. Water Resources Research 5:1070-1077.

Pickard, W. F. 1981. The ascent of sap in plants. Progress in Biophysical Molecular Biology 37:181-229.

Proffitt, A. P. B., P. R. Berliner, and D. M. Oosterhuis. 1985. A comparative study of root distribution and water extraction efficiency by wheat grown under high- and low-frequency irrigation. Agronomy Journal 77:655-662.

Reid, J. B. and M. G. Huck. 1990. Diurnal variation of crop hydraulic resistance: A new analysis. Agronomy Journal 82: (In Press).

Reid, C. P. 1985. Mycorrhizae: A root-soil interface in plant nutrition. In Microbial-plant interactions, ed. R. L. Tood and J. E. Giddens, 29-50. Madison, WI: ASA Special Publication no. 47. American Society of Agronomy.

Reid, J. B., and B. Hutchison. 1986. Soil and plant resistances to water uptake by Vicia faba L. Plant and Soil 92:431-441.

Rogers, H. H., and P. A. Bottomley. 1987. In-situ nuclear magnetic resonance imaging of roots: Influence of soil type, ferromagnetic particle content, and soil water. Agronomy Journal 79:957-965.

Ruehle, J. L. 1983. The relationship between lateral root development and spread of Pisolithus tinctorius ectomycorrhizae after planting of container grown loblolly pine seedlings. Forest Science 29:519-526.

Schulze, E.-D., R. H. Robichaux, J. Grace, P. W. Rundel, and J. R. Ehleringer. 1987. Plant water balance: In diverse habitats, where water often is scarce, plants display a variety of mechanisms for managing this essential resource. BioScience 37:30-37.

Soil Survey Staff. 1975. Soil taxonomy. A basic system of soil classification for making and interpreting soil surveys. Agriculture Handbook no. 436. Washington, D.C.: Soil Conservation Service. U.S. Department of Agriculture, Soil Conservation Service.

Stoker, R., and P. E. Weatherley. 1971. The influence of the root system on the relationship between the rate of transpiration and depression of leaf water potential. New Phytologist 70:547-554.

Swartzendruber, D., and D. Hillel. 1973. The physics of infiltration. In Physical aspects of soil water and salts in ecosystems, ed. Hadas et al., 3-15. New York: Springer-Verlag.

Taylor, H. M. 1983. Managing root systems for efficient water use: An overview. In Limitations to efficient water use in crop production, ed. H. M. Taylor, W. R. Jordan, and T. R. Sinclair, 87-113. Madison, WI: American Society of Agronomy.

Taylor, H. M., M. G. Huck, and B. Klepper. 1972. Root development in relation to soil physical conditions. In Optimizing the soil physical environment toward greater crop yields, ed. D. Hillel, 57-77. New York: Academic Press.

Taylor, H. M., and B. Klepper. 1975. Water uptake by cotton root systems: An examination of assumptions in the single root model. Soil Science 120:57-67.

Taylor, H. M., and B. Klepper. 1978. The role of rooting characteristics in the supply of water to plants. Advances in Agronomy 30:99128.

Taylor, H. M., B. Klepper, and R. W. Rickman. 1979. Some problems in modeling root water uptake. ASAE Paper no. 79-4516. St. Joseph, MI: American Society of Agricultural Engineers.

Thompson, P. J., I. J. Jansen, and C. L. Hooks. 1987. Penetrometer resistance and bulk density as parameters for predicting root system performance in mine soils. Soil Science Society of America Journal 51:1288-1293.

Tinker, P. B. 1976. Roots and water. Transport of water to plant roots in soil. Philosophical Transactions of the Royal Society of London B. 273:445-461.

312

Van Bavel, C. H. M., and J. M. Baker. 1985. Water transfer by plant roots from wet to dry soil. Naturwissenschaften 72:606-607.

Van Bavel, C. H. M., K. J. Brust, and G. B. Stirk. 1968. Hydraulic properties of a clay loam soil and the field measurement of water uptake by roots: II. The water balance of the root zone. Soil Science Society of America Proceedings 32:317-321.

Van Bavel, C. H. M., G. B. Stirk, and K. J. Brust. 1968. Hydraulic properties of a clay loam soil and the field measurement of water uptake by roots: I. Interpretation of water content and pressure profiles. Soil Science Society of America Proceedings 32:310-317.

Van Genuchten, M. Th., and P. J. Wierenga. 1976. Mass transfer studies in sorbing porous media. I. Analytical solutions. Soil Science Society of America Journal 40:473-480.

Van Noordwijk, M., and P. de Willigen. 1987. Agricultural concepts of roots: from morphogenetic to functional equilibrium between root and shoot growth. Netherlands Journal of Agricultural Science 35:487-496.

Wang, J., J. D. Hesketh, and J. T. Woolley. 1986. Preexisting channels and soybean rooting patterns. Soil Science 141:432-437.

Weatherley, P. E. 1975. Water relations of the root system. In The development and function of roots, ed. J. G. Torrey and D. T. Clarkson, 397-413. New York: Academic Press.

Weatherley, P. E. 1976. Introduction: Water movement through plants. Philosophical Transactions of the Royal Society of London B. 273:435-444.

Weatherley, P. E. 1982. Water uptake and flow in roots. In Physiological plant ecology II, ed. O. L. Lange, P. S. Nobel, C. B. Osmond, and H. Ziegler, 79-109. Encyclopedia of Plant Physiology. n.s., volume 12B, Berlin: Springer-Verlag.

Willatt, S. T., and H. M. Taylor. 1978. Water uptake by soya-bean roots as affected by their death and soil water content. Journal of Agricultural Science 90:205-213.

Wilun, Z., and K. Starzewski. 1972. Soil Mechanics in Foundation Engineering. New York: John Wiley & Sons.

Zur, B., J. W. Jones, K. J. Boote, and L. C. Hammond. 1982. Total resistance to water flow in field soybeans: II. Limiting soil moisture. Agronomy Journal 74:99-105.

INDEX

Adaptive strategies, 278
Aeration, 219, 279
Agroecosystems, 3
Apical meristem, 262
Azospirillum, 120-121, 256
  brasilense, 120
  chroococcum, 120
  lipoferum, 120-121
  paspali, 121

Bacillus, 119
Beijerinckia, 119
Biomass, 234-235, 237,
  239-242, 245
Bradyrhizobium, 123
Bulk density, 270
Burrow, 193-195, 197-201,
  205, 207-209

Capillary
  flow, 277-278
  movement, 282
Carbohydrate partitioning,
  286
Carbon, 47, 51, 54-55,
  58-61, 63-64, 66, 151,
  156, 194, 242
Carbon balance, 286
Casts, 196-199, 203, 205,
  207-208
Catenary, 281
Clinostat, 252

Cohesion, 281
Collembola, 100, 102-112,
  117-121, 149
Continuum, plant-soil-
  atmosphere, 285
Copper, 148

Drainage, 268, 274, 278-279,
  286, 300
  channels, 273
Drought stress, 278, 282,
  289, 300

Earthworm, 192-194, 196-199,
  201-209
Ectorhizosphere, 2
Endodermis, 283
Endorhizosphere, 2, 119
Enterobacter, 117
Epidermis, root, 283-284
Erosion, 270
Ethylene, 262-264
Evaporation, 268, 277-279,
  283, 286, 294, 297, 300
Exorhizosphere, 119
Exudate, 58-61

Fauna, 149
Forrester diagram, 286
Fluid flow, 276
Flux, 291
  density, 277

313

Friable, 269

Gaseous diffusion, 271
Gravitational potential,
274
Gravity, 83, 252-253,
256-257
Growth, 268-270, 274,
277-280, 286, 294, 297,
300, 302
rates, 285, 289, 291
Growth substances, 262, 264
Guttation, 281

Horizontal
flow, 282
layers, 286-287
Humus, 50-51, 54
Hydraulic
conductivity, 277-278,
282
potential, 276-277
properties, 268, 270
Hydrophilic, 271
Hyphae, 144-152, 154-159,
284

Indole-3-acetic acid (IAA),
130, 262
Infiltration, 194, 202,
207-208, 269, 273, 286
Internal water balance, 286
Irrigation, 268, 270,
273-274, 279, 286, 297,
300, 302

Lignification, 284

Matric potential, 274
Mechanical
force, 270
impedance, 262-264

Mechanical (con't)
resistance, 279
support medium, 269
Mesofauna, 2-3, 5
Microgravity, 252-257
Minirhizotron, 100, 102-104,
110, 117, 119, 120-121
Moldboard plow, 12-13, 34,
37
Monoliths, 88, 91, 95
Mucigel, 255-256
Mulch, 277
Mycelia, 284
Mycorrhiza, 2-3, 59-60,
144-145, 147, 151-152,
154, 156-159, 284

Nitrogen, 47-49, 51-56,
58-59, 61-66, 217,
219-220
Nitrogen fixation, 117,
119-122, 128-131,
133-134
associative, 119
symbiotic, 122
Nodulation, 128, 130, 131
Non-irrigated
plants, 290
root systems, 300
Organic matter, 23, 33,
47-49, 51, 54, 58,
60-61, 63-64, 148-149,
193-194, 196, 198, 205,
271
Osmotic potential, 279-280
Outer space, 252-253, 256

Packing density, 271, 276
Particle size, 271
Percolate, 268-269, 274
pH, 52, 54, 57-58, 123,
125, 128, 148, 196,

pH (con't) 206-207
Phosphorus, 51, 144,
    147-149, 154-156,
    196-197, 200, 203
Photoassimilate, 217, 219,
    221, 226, 228-229
Photoperiod, 221, 223-225
Photosynthate, 2, 59, 84,
    232-234, 240, 244
Photosynthesis, 282, 286
Phytochrome, 225
Plant water potential, 286,
    289, 291
Polysaccharides, 117,
    133-134, 157
  pore spaces, 270-271,
    274, 276, 279
  soil, 232, 274
Porosity, 202, 269, 277
  soil, 270
Porous media, 276
Potential gradient, 277,
    281-283, 297

Rainfall, 268, 270, 273-274,
    277-280, 286, 291, 294,
    296, 300
Regrowth, 279
Resistance, 271, 278, 281,
    282-284, 294, 297, 300
Rhizoplane, 119
Rhizosphere, 2-3, 5, 16,
    47-48, 51-52, 57-64, 66,
    83-85, 102-104, 109,
    117, 119, 144, 149, 192
Rhizotron, 98, 100, 110
Root, 100, 103-104, 113-121,
    144-152, 154, 156-158,
    192, 194-203, 206-209,
    217-219, 221-225, 232,
    235, 237-240, 268, 270,

Root (con't) 274, 277-286,
    289, 291, 294, 297, 300
  cap, 254-255, 262, 264
  crops, 229
  development, 10, 15, 29,
    38, 42, 63-65, 263
  distribution, 244
  distribution profile, 295
  elongation, 84, 107, 253,
    262
  growth, 242-243, 245,
    252-253, 255, 262, 265,
    269-270, 274, 277-280,
    285-286, 289, 291,
    293-294, 297, 300, 302
  hairs, 284
  length, 35, 38, 40, 42-43,
    234-237, 241, 278, 281,
    284, 289, 291-292, 294
  systems, 29, 83-89, 91-92,
    95-98, 103-105, 107,
    109-110, 226, 228, 232,
    234-235, 237, 245-246
  tips, 279
  washing, 97
  zone, 10, 15-18, 21, 25,
    28
Root/shoot, 38, 84, 237-240,
    243-245
Runoff, 268

Salinity, 119, 122-123, 125,
    128
Seepage, 273
Shoot, 232, 234, 238, 240,
    242-246
  elongation, 253
Shoot/root, 217-218,
    220-221, 225-226,
    228-229
Simulation, 285, 289, 291,

Simulation (con't) 300
Soil
  aeration, 26, 28, 31, 33,
    36-39, 51, 207-208
  compaction, 8, 10-11,
    13-14, 28-31, 33, 37,
    64, 208
  drainage, 207, 209
  gaseous flux, 43
  heat flux, 21, 23-24, 64
  macroporosity, 10-11, 33
  particles, 268-272, 274,
    283
  porosity, 10-11, 15-16,
    21, 23-24, 27-28,
    31-33, 37, 201
  strength, 13, 31, 33, 36,
    38-39, 43, 243
  structure, 8, 10, 12, 15,
    18, 29, 30, 34, 42-44,
    63-64, 198, 203
  subsoil, 280, 300
  temperature, 16-17, 20,
    24-27, 34, 39, 42, 47,
    51, 58, 64, 83, 85,
    107, 123, 125, 128, 194
  water, 47, 49, 51, 57-58,
    61, 63-64, 66, 194,
    196-199, 203, 207-208,
    217-220, 229, 232-233,
    243-244, 268, 270,
    279-280, 285, 289, 294,
    300
  water content, 270,
    277-278, 286, 289, 291,
    297, 300, 302
  water flux, 10, 24, 28,
    30
  water potential, 282,
    285, 291, 294, 297
  water storage, 268, 269
Stele, 283

Stomatal
  cavities, 281
  closure, 281-282, 289, 300
Surface drainage, 273
Swamp, 274
System boundaries, 286-287

Temperature, 221, 223, 228,
  233, 242-243
Tillage, 9-15, 20-26, 29,
  32-33, 37, 39, 42, 47,
  49, 63-65, 192, 194-195,
  201, 204-206, 208-209
Transpiration, 278, 281-282,
  286, 289, 297
Trichoderma, 117

Unsaturated
  flow, 276, 282, 294
  hydraulic conductivity,
    278
  soil, 269, 277

Vertical, 301
Vapor phases, 270, 273
Vertical, 282, 294, 297

Water, 144, 148-149, 157
  content, 270-271, 277-278,
    281, 286, 297, 300
  extraction, 283, 297, 300
  movement, 270, 272, 274,
    277, 281-283, 294, 300
  potential, 268, 277,
    281-282, 285-286, 289,
    291, 294, 297, 300
  table, 274, 277, 299
  uptake rate, 281-282, 291,
    297

Xylem, 278, 281, 283-284,
  300

ABOUT THE CONTRIBUTORS

<u>Raymond R. Allmaras</u> is a Soil Scientist with the Agricultural Research Service, U.S. Department of Agriculture, in St. Paul, Minnesota. His areas of expertise include soil physics, the plant-soil-water-air system, soil management, and soil and water conservation. Dr. Allmaras has published extensively on root growth and the field environment related to tillage and residue management, root disease interactions with the soil physical environment, and soil physical and chemical properties related to soil management of tillage, traffic, and crop residues. He received his B.S. degree from North Dakota State University, his M.S. from the University of Nebraska, and his Ph.D. from Iowa State.

<u>James E. Box, Jr.</u> is a Soil Scientist with the Southern Piedmont Conservation Research Laboratory of the U.S. Department of Agriculture's Agricultural Research Service in Watkinsville, Georgia. He was Laboratory Director of the Center (1966-1984), served on the Editorial Board of the <u>Journal of Soil and Water Conservation</u> (1981-1989), and was Chairman of the Southern Regional Information Exchange Group for Root Environment (1986-1989). He has authored and co-authored more than 90 scientific publications, and participated in international symposia on root research. Dr. Box's particular areas of interest are use of the minirhizotron technique and rooting of soft red winter wheat. He received his B.S. and M.S. degrees from Texas A&M University and his Ph.D. from Utah State University.

<u>C. E. Clapp</u> is a Research Chemist with the Soil and Water Management Unit, USDA-Agricultural Research Service, and Professor of Soil Biochemistry at the University of

317

Minnesota, St. Paul. He has authored chapters in <u>Sewage Sludge Organic Matter</u> (1986), <u>Viscosity of Humic Substances</u> (1989), <u>Adsorption of Organics on Soil Colloids</u> (1990), and <u>Humic Substances</u> (1990). The chemistry of soil organic matter, nitrogen transformations, clay-organic complexes, water quality, and municipal wastes are among Dr. Clapp's special areas of expertise. He received his B.S. in chemistry from the University of Massachusetts, and his M.S. and Ph.D. degrees in soil chemistry from Cornell University.

<u>Robert H. Dowdy</u> is a Research Leader with the Agricultural Research Service, U.S. Department of Agriculture, St. Paul, Minnesota. His multidiscipline research unit is concerned with the impact of soil tillage on soil structure, root development, water flow, energy fluxes, residue decomposition, and processes controlling the movement of pesticides and nitrates through soils. His personal research deals with plant root development as mediated by tillage and residue management. Dr. Dowdy won first place in the U.S. Environmental Protection Agency's 1989 National Beneficial Sludge Use Award. He received his B.S. degree from Berea College, his M.S. from the University of Kentucky, and his Ph.D. from Michigan State University.

<u>Lewis J. Feldman</u> is an Associate Professor of Plant Biology at the University of California, Berkeley. His areas of specialization include root development and gravitropism. Dr. Feldman received his Ph.D. from Harvard University.

<u>Luther C. Hammond</u> is Professor of Soil Physics at the University of Florida, Gainesville. His multidiscipline investigations includes water retaining and transmitting characteristics of the major soils of Florida; plant water use and management in response to crop and soil management variables and to atmospheric demand; root growth in relation to plant and soil factors; and spatial variability of soil physical properties and crop yields. He received his B.S. from Clemson University and his M.S. and Ph.D. from Iowa State University.

<u>Gerrit Hoogenboom</u> is Assistant Professor, Department of Agricultural Engineering, University of Georgia, Georgia Experiment Station, Griffin. His particular areas of expertise are crop physiology and crop modeling. He co-authored <u>Advances in Irrigation 4</u> (1987) and has authored/co-authored many refereed articles and monographs. Dr.

Hoogenboom received his B.Sc. and M.Sc. degrees from Agricultural University, Wageningen, the Netherlands, majoring in horticulture/crop modeling, and his Ph.D. from Auburn University, with a major in crop science.

Morris G. Huck is a Soil Scientist with the Agricultural Research Service, U.S. Department of Agriculture, in Urbana, Illinois. Root physiology, root/soil interactions, and computer simulation of biological processes at the organism/field scale are his specific areas of specialization. Dr. Huck received his B.S. and M.S. degrees from the University of Illinois at Urbana and his Ph.D. from Michigan State.

P. G. Hunt is Research Leader/Director of the USDA-Agricultural Research Service's Coastal Plains Research Center in Florence, South Carolina. He was co-recipient of the L.M. Ware Award for distinguished research, American Society of Horticulture, Southern Region, in 1990, and served as associate editor of the Journal of Environmental Quality from 1976 to 1978. Soil microbiology -- particularly nitrogen cycling and land treatment of waste -- is Dr. Hunt's area of specialization. He received his B.S. and M.S. degrees in agronomy from Clemson University and his Ph.D. from the University of Florida.

Michael J. Kasperbauer is a Research Plant Pathologist at the Coastal Plains Research Center, Agricultural Research Service, U.S. Department of Agriculture, at Florence, South Carolina, and Professor (Adjunct) of Plant Pathology at Clemson University. His career research is primarily on the physiology of plant growth and development and the use of cell and tissue culture in plant improvement. Dr. Kasperbauer has authored or co-authored more than 150 research papers. He served as associate editor of Agronomy Journal from 1975 to 1983, and edited/co-authored Biotechnology in Tall Fescue Improvement. He received his Ph.D. degree from Iowa State University.

Eileen J. Kladivko is an Associate Professor of Agronomy at Purdue University, West Lafayette, Indiana. Within the area of soil physics, her particular research topics are conservation tillage, earthworms in agricultural systems, and water quality. Dr. Kladivko has spoken to many farm groups about earthworms and agricultural management. She received B.S. and M.S. degrees from Purdue in environmental science and agronomy, respectively, and a Ph.D. in soil science from the University of Wisconsin-Madison.

S. D. Logsdon is a Soil Scientist with the Agricultural Research Service, U.S. Department of Agriculture, at St. Paul, Minnesota. She has published articles on root growth in aggregated soil and as influenced by soil physical properties. Her particular areas of specialization are applied soil physics and soil-plant relations. Dr. Logsdon received her B.A. from Cedarville College, her M.S. from Michigan State University, and her Ph.D. from Virginia Polytechnic Institute and State University.

B. L. McMichael is a Plant Physiologist with the U.S. Department of Agriculture-Agricultural Research Service Cropping Systems Research Laboratory in Lubbock, Texas. His particular areas of expertise include the physiology of plant root systems and the impact on productivity of root system genetic diversity. He has authored Growth of Roots in Cotton Physiology (1986) and journal articles on the occurrence of different vascular systems in cotton roots and their relation to lateral root development. Dr. McMichael serves as vice-president and co-editor of Proceedings of the International Society of Root Research. He received his B.S., M.S., and Ph.D. degrees from Texas A&M University in agronomy, soil physics, and plant physiology, respectively.

J. A. E. Molina is Professor of Soil Microbiology at the University of Minnesota, St. Paul. He has authored/co-authored numerous refereed journal articles during the 1980s, with particular focus on carbon and nitrogen flux in the soil-plant system. Dr. Molina's undergraduate education consists of a Baccalaureate degree from the University of Paris and graduation from the Institut National Agronomique in Paris. He majored in agronomy-soil microbiology for his M.S. and Ph.D. degrees from Cornell University.

Randy Moore is a Professor in and Chairman of the Department of Biological Sciences, Wright State University, Dayton, Ohio. He has published extensively, as well as editing Vegetative Compatibility Responses in Plants (1983), CRC Handbook of Plant Cytochemistry (1987), and Laboratory Exercises in Biology (1986). Dr. Moore has served as editor-in-chief of American Biology Teacher since 1985, editor of The Desktop Biologist since 1987, and was formerly editor-in-chief of Journal of the Texas Society for Electron

Microscopy and associate editor of Texas Journal of Science. He served as a Fulbright Scholar to Thailand in 1987. Dr. Moore received a B.S. in biology from Texas A&M University, an M.S. in botany from the University of Georgia, and a Ph.D. in the field of plant development from the University of California, Los Angeles.

Alvin J. M. Smucker is Professor of Soil Biophysics, Department of Crop and Soil Sciences, Michigan State University, East Lansing. Dr. Smucker has patented a hydropneumatic elutriation system and is coordinator of a global root imaging network. His special areas of expertise include image processing of plant root systems, root carbon utilization, and root morphological and physiological responses to stress environments. He received his B.S. in biology from Goshen College and his M.S. and Ph.D. degrees in soil science and soil physics, respectively, from Michigan State.

Renate Machan Snider is Assistant Professor in the Zoology Department, Michigan State University, East Lansing. She is currently investigating the effect of extremely low frequency electromagnetic fields on earthworms and soil arthropods (Project ELF in Michigan's Upper Peninsula) and working on the systematics and distribution of millipedes at the Savannah River Site, South Carolina, and on the island of Bermuda. Dr. Snider received her B.A. from the University of Vienna and her Ph.D. from Michigan State.

Richard J. Snider is a Professor and Director of Undergraduate Programs, Zoology Department, Michigan State University, East Lansing. He has authored chapters in An Introduction to the Aquatic Insects (1984) and Immature Insects (1987). Dr. Snider's specific areas of specialization are the bionomics and systematics of Collembola and the effects of tillage practices on soil arthropod populations. He is currently president of the Michigan Entomological Society. He received his B.S., M.S., and Ph.D. degrees from Michigan State.

David M. Sylvia is an Associate Professor in the Soil Science Department, University of Florida, Gainesville. He is currently associate editor for the Soil Science Society of America journal, and served as a National Academy of Sciences Exchange Scholar to Czechoslovakia in 1989. His present research interests include ecology and physiology of mycorrhizae and rhizosphere biology. During the last five years, Dr. Sylvia has authored/co-authored numerous refer-

eed journal articles, two monographs, and a chapter in
Applied Mycology (in press). He received his B.S. in for-
estry and M.S. in plant pathology from the University of
Massachusetts, Amherst, and his Ph.D., with a major in
plant pathology, from Cornell University.

Howard M. Taylor is Rockwell Professor of Agronomy, Horticulture, and Entomology at Texas Tech University, Lubbock.
He served as co-author of _Principles of Soil-Plant Interre-_
lationships (1989) and editor/co-editor of _Minirhizotron_
Observation Tubes and _Limitations to Efficient Water Use_
for the American Society of Agronomy and _Modifying the Root_
Environment and _Compaction of Agricultural Soils_ for the
American Society of Agricultural Engineers. Dr. Taylor has
lectured on root growth and function in approximately 25
countries. He received his B.S. from Texas Tech and his
Ph.D. from the University of California, Davis.

Hubert J. Timmenga is with the Department of Soil Science,
University of British Columbia, Vancouver, Canada. He has
authored chapters in _Earthworm Ecology_ (1983), _Earthworms_
and Heavy Metals, and _Soil Biology and Biochemistry_ (1986).
Dr. Timmenga's particular areas of interest include earthworm ecology, waste management, and composting. He received his M.S. from Agricultural University, Wageningen,
The Netherlands, and his Ph.D. in soil science from the
University of British Columbia.

Dan R. Upchurch is a Soil Physicist with the USDA-
Agricultural Research Service's Plant Stress and Water
Conservation Unit in Lubbock, Texas. His current research
program is focused on understanding the limitations to
water flow in the soil-plant-atmosphere continuum from both
physical and physiological views, in order to determine the
optimum root system characteristics for water-limited
environments. Dr. Upchurch has published extensively in
technical journals. He received his B.S. from New Mexico
State University, majoring in agronomy (soils), his M.S.
from the University of California, Davis, majoring in soil
science, and his Ph.D. from Texas Tech University.

M. C. Whalen is an Assistant Professor in the Department of
Biology, Colby College in Waterville, Maine. Her research
interests encompass the molecular genetics of plant-microbe
interactions and plant development. Dr. Whalen received
her Ph.D. in botany from University of California at
Berkeley.